Aspiring Academics

A Resource Book for Graduate Students and Early Career Faculty

EDITORS

Michael Solem, *Association of American Geographers*
Kenneth Foote, *University of Colorado at Boulder*
Janice Monk, *University of Arizona*

AAG
ASSOCIATION *of*
AMERICAN GEOGRAPHERS

PEARSON
Prentice
Hall

Upper Saddle River, NJ 07458

Library of Congress Cataloging-in-Publication Data
Aspiring academics : a resource book for graduate students and early career faculty / editors,
Michael Solem, Kenneth Foote, Janice Monk.
    p. cm.
 Includes bibliographical references and index.
 ISBN-13: 978-0-13-604891-6 (alk. paper)
 ISBN-10: 0-13-604891-9 (alk. paper)
 1. Graduate students--Employment--United States. 2. College teachers--Vocational guidance--
United States. 3. Job hunting--United States.   I. Solem, Michael. II. Foote, Kenneth. III. Monk,
Janice.
 LB2332.72.A87 2009
 378.1'2023--dc22

                                     2008003311

Editor-in-Chief, Science: Nicole Folchetti
Publisher, Geosciences and Environment: Dan Kaveney
Project Manager: Tim Flem
Marketing Manager: Amy Porubsky
Production Manager: Kathy Sleys
Creative Director: Jayne Conte
Cover Design: Maureen Eide
Cover Illustration/Photo: Nicole Anderson, Fred Shelley, and Michael Solem
Full-Service Project Management/Composition: Karpagam Jagadeesan/GGS Book Services
Printer/Binder: Courier Corporation/Stoughton

This material is based upon work supported by the National Science Foundation under
Grant No. REC-0439914 and DUE 0089434. Any opinions, findings and conclusions, or
recommendations expressed in this material are those of the author(s) and do not necessarily
reflect the views of the National Science Foundation (NSF).

Credits and acknowledgments borrowed from other sources and reproduced, with permission,
in this textbook appear on appropriate page within text.

Pearson Education Ltd., London
Pearson Education Singapore, Pte. Ltd
Pearson Education Canada, Inc.
Pearson Education –Japan
Pearson Education Australia PTY, Limited
Pearson Education North Asia, Ltd., Hong Kong
Pearson Educación de Mexico, S.A. de C.V.
Pearson Education Malaysia, Pte. Ltd.
Pearson Education Upper Saddle River, New Jersey

10 9 8 7 6 5 4 3 2 1
ISBN-13: 978-0-13-604891-6
ISBN-10:   0-13-604891-9

CONTENTS

Aspiring Academics and Teaching College Geography: A Community-Based Web Site for New Scholars

The companion web site for this book (www.pearsonhighered.com/aag) contains a wealth of additional resources for professional development and community building.

Each chapter is supported by activities for workshops, courses, seminars, brown bags, and informal gatherings among graduate students and faculty. The activities provide procedures, recommended readings, worksheets, and other supplementary materials that can help you get started with applying new skills and ideas in your own professional practice.

The web site also features tools for sharing ideas and effective practices with others. Use the discussion boards to exchange perspectives and advice about professional development issues with the larger community. Alternatively, you can set up private forums for groups organized by teaching assignment, academic department, research network, or other interest area.

You can also participate in the development of the web site by contributing your own classroom activities, links, and professional development resources.

We hope you find the web site to be a valuable complement to the text.

PREFACE

Creating a Climate of Support for Graduate Students and Early Career Faculty

A Call for Change and Action

Michael Solem, Kenneth Foote, and Janice Monk

OVERVIEW

This book and its companion *Teaching College Geography* (Solem and Foote 2009) are for people beginning their careers in higher education. Our aim is to focus on topics that are important to success in an academic career and that are frequently the source of the greatest stress in first years of a college or university appointment.

A major reason for this stress is that, though our graduate programs provide first-rate training in research theory and methods, and usually some level of teaching apprenticeship, very few of those programs address some of the key professional responsibilities that most early career faculty will face in their first years of a full- or part-time appointment—responsibilities such as preparing and leading a complete course as primary instructor; working effectively with colleagues; dealing with nuts-and-bolts issues of writing and publishing such as responding effectively to peer reviews and editors' comments; finding and then managing grant funding; managing time effectively; confronting ethical issues in research and teaching; balancing work, family, and personal interests and responsibilities; creating a professional development plan; and many other issues.

This situation—in which many key faculty responsibilities and many of the everyday duties of academic life are not addressed in graduate curricula—is not uncommon in doctoral programs in the social and natural sciences (Golde and Dore 2004; Golde and Walker 2006). In fields such as law and medicine, internship programs, clinical rotations, and residencies help lawyers and doctors make smooth transitions between classroom theory and professional practice. For academics, this transition tends, instead, to be a "sink or swim" experience. New hires who are able to identify and quickly meet the expectations for their work are much more likely to succeed than those who cannot. Many graduate programs, departments, and universities do, of course, take a different and more supportive approach by offering mentoring programs, help on issues of learning and teaching, and regular professional development opportunities for graduate students and early career faculty. Nationally recognized efforts such as the Preparing Future Faculty program provide another model for effective professional development (Gaff, Pruitt-Logan, and Weibl 2000; Wulff and Austin 2004).

Mentoring and other forms of professional support increase the chance of success for new academics, with a number of other benefits. Reducing needless stress and providing a clearer idea of expectations means early career faculty can contribute more fully to department life and be more productive in their work from the very start of their appointment. Even self-starters who already know how to swim can benefit from timely support because, as research shows, help provided in the first few semesters of a new appointment seems to have a positive influence on attitudes and work well beyond tenure (Fink 1984; Sorcinelli and Austin 1992; Boice 1992, 2000). Higher levels of success in retention and promotion do have financial value for most colleges and universities when judged against the resources invested in recruiting and supporting new hires through the pre-tenure years. But beyond any financial considerations, faculty who find the support to succeed are likelier to be more engaged in and contribute more to an institution's mission than those who leave. Timely and ongoing support has the added benefit of making the expectations of academic life clearer to all early career faculty. Without this clarity and openness, issues of key importance to professional success remain hidden and privileged to the detriment of early career faculty who by reason of personal background, graduate training, gender, nationality, sexuality, family status, age, disability, or other characteristics may be excluded from discussion of these issues or be hesitant to ask about them. We see improved professional development as a key to diversifying the professoriate to better reflect today's demographic and intellectual realities by opening higher education to the best scholars and scientists irrespective of their backgrounds and supporting them throughout their careers.

OUR VISION OF PROFESSIONAL DEVELOPMENT

To make a difference to early career faculty, we believe that professional development is more than simply offering an occasional workshop, establishing a mentoring program for new hires, or introducing the topics of this book into graduate workshops. We know that such plans tend to be more effective when they are proactive rather than reactive, long-term rather than short-term, and have a broad base of support within a department, college, or university; but their impact will be heightened if they are developed as interlocking elements of a broader plan of professional development.

Our Vision for This Book is Based on the Following Principles

First, it is important to view the many responsibilities of academic life together, in balance, and to understand how teaching, research, service, outreach, and our personal lives intersect and interconnect. Many self-mentoring books for early career faculty focus on writing, publishing, or teaching alone. We are attempting to take a broader, integrated approach in *Aspiring Academics*, one aligned with Boice's vision of moderation, balance, and the seamless, positive integration of the diverse elements of life and work. We stress issues of teaching and service that are so often of paramount importance to academic life, but not to the exclusion of topics such as time management, work–life balance, career planning, and collegiality. We believe these topics are, together, the keys to long-term success and career satisfaction. By considering them together it is possible to make choices among short-term and long-term goals and recognize potential conflicts before they arise.

Second, achieving this balance involves adopting the conception of teaching as scholarly work as proposed by Ernest Boyer in *Scholarship Reconsidered* (1990) and since developed into a broad-based, international movement to promote the scholarship of teaching and learning (SoTL) (Glassick, Huber, and Maeroff 1997). SoTL strives to promote significant, long-lasting learning for all students as well as to reward and recognize effective teaching in the same way as other types of scholarly work do. SoTL implies that we apply the same standards of rigor we use in our scholarship and scientific research to our approaches to teaching and learning. In part, this means opening up our teaching for discussion, feedback, review, and collaboration among colleagues, peers, and students—just as we would do with our research—rather than seeing teaching as something private and confined to the classroom. This also means that we can strive to create more effective linkages between our teaching and research and, indeed, actually researching our teaching, teaching materials, and teaching strategies to see whether they are effective—and disseminating the results in print and in public presentations. Most importantly, SoTL implies that we become critically reflective teachers who constantly consider the effectiveness of our practices, keep abreast of research on teaching and learning, and contribute to this process of improvement.

Third, as SoTL implies, effective professional development is not just sharing war stories about teaching in the trenches, but instead must be anchored in a solid knowledge of relevant research and literature on learning theory, curriculum design, student assessment, curriculum alignment, project evaluation, instructional technology, and other topics. The fact is that educational psychologists, curriculum developers, and others understand much about how people learn and how to undertake effective course and curriculum planning. Yet, as a consequence of our current process of graduate education, few early career faculty will be aware of more than a few limited concepts from these literatures. Instead, most of us get started by simply using the techniques—effective or ineffective—to which we were exposed as students. The authors of the following chapters have made every effort to situate their discussions in current research, and all of the chapters in this book contain extensive reference lists that, we hope, will push readers to expand their understanding of essential background literatures.

Fourth, professional development must be seen as a lifelong, career-spanning effort, not something confined to graduate school and the first years beyond. In this book, we focus on the early career period, but only because previous research mentioned above has indicated how critical this period is in building career momentum. We concur with Healey (2003), however, that more attention should be devoted to issues of professional development from graduate school onward. There is a range of topics relevant to different academic career stages and a "just in time," individualized approach to providing effective, timely training is better than trying to pack all these topics into graduate school and the early career years. Leadership training for chairing a department will, for example, be of more interest to senior faculty, but only for some senior faculty at particular steps in their careers. At the same time, we believe that many of the following chapters will be of as much interest to senior professors as to early career faculty. Since many of these topics are rarely part of graduate curricula, senior faculty will not necessarily have the background needed to introduce them in graduate seminars. For this reason, the chapter authors have developed activities that can be used in seminars, workshops, and other settings "as is" or with minor revisions. All of these activities can be accessed and downloaded from this book's web site.

Fifth, discipline-based efforts are as important as interdisciplinary professional development programs typically available on campus (Healey and Jenkins 2003). Many colleges and universities provide excellent programs, workshops, and seminars on career topics for graduate students and early career faculty, such as the Preparing Future Faculty Program. Other initiatives focus on particular themes (such as promoting SoTL or service learning, for instance), and still others such as the National Science Foundation's ADVANCE grants aim to provide leadership training for women and underrepresented groups. However, these sorts of campus-wide efforts must also be complemented with crosscutting discipline-specific professional development opportunities. There are special teaching challenges in most disciplines that need to be raised and, as Becher (1989) has argued, there are cultural and social differences among disciplines, which prevent some sharing of insights across disciplines. Although this book stems from a discipline-wide professional development effort in geography, our intent is that readers should also take advantage of interdisciplinary programs whenever and wherever possible.

Finally, networking and mentoring are essential to early career development. Or perhaps we should say that professional development should be personal, friendly, and a social part of our day-to-day lives. Too often early career faculty report feelings of isolation and having to "go it alone" in their first few years without recognizing that others are facing similar stresses. Too often, individuals internalize their anxieties thinking they have to find the solutions on their own when, in fact, others are more than willing to help. Particularly important are mentoring and networking opportunities, which extend beyond one's own graduate school cohort and home department. Too frequently, individuals are reticent about asking seemingly obvious questions among their department colleagues feeling they will lose face or possibly be judged unfavorably for merit or promotion. Building a more extensive network, for example through contacts at professional meetings, is important to dispelling these feelings. This book cannot by itself help with this networking, but we do encourage readers to participate in national and regional meetings, to organize paper and panel sessions, and to contact scholars in their field to develop intellectual and personal friendships. Hardwick's (2005) article on mentoring strategies as well as Moss et al. (1999) on self-mentoring strategies provide many good ideas.

QUESTIONING OUR ACADEMIC CULTURE AND CLIMATE: A MANIFESTO FOR CHANGE

At a deeper level, our concern to promote change in the culture of academic life is to better support early career faculty. Too often, it seems that the burden of professional development is placed on individual faculty—it is their responsibility alone to improve their writing, strengthen their teaching, and manage their time effectively. Indeed, much of the research into professional socialization has focused largely on personal attributes and individual behaviors. Boice (1992, 2000), for instance, has found that early patterns of success among new faculty "quick starters" are closely related to efficient and strategic time management strategies, which help the new professor effectively balance teaching, research, and service responsibilities with the demands of personal life. Boice further notes that a successful beginning in the tenure-track career path is associated with being proactive and seeking feedback from colleagues, becoming familiar with local institutional culture, and learning the literature informing educational practice in one's discipline. The chapters in this book on time management and collegiality reinforce Boice's ideas, yet they also go further by pointing out the responsibilities of the overall departmental community and roles that a collective of faculty and graduates can play in enhancing the actions taken by individuals to develop as academic professionals.

We believe, moreover, that this "self-help" view of professional development must change to acknowledge the influence of the academic and social environments of particular departments, colleges, universities, and communities. Effort directed toward self-improvement is important but will also be constrained or enhanced by the institutional climate and culture (Lee 2007). This point is especially salient in the context of an emerging literature showing evidence that variables from student completion rates to faculty productivity are sensitive to factors "that are not simply a function of personal attributes" (Council of Graduate Schools 2003, 11). In other words, even the careers of the most talented and gifted academics are vulnerable to the consequences of infighting, territoriality, competition, and the biases of colleagues who are dismissive of talented scholars because of their gender, nationality, minority status, or other characteristics unrelated to professional ability and accomplishment. All of these factors can degrade a department or university's climate and ultimately cause many faculty to leave, just as supportive colleagues can create a climate where early career faculty are more likely to thrive (Cameron and Ettington 1988; Petersen and Spencer 1990; Rosser 2004).

We acknowledge that academic departments have cultures distinguishing them from other professional workplaces, but we also wish to account for the roles that various human actors play in shaping those cultures and who, over time, possibly contribute to normative shifts within departments. Just as academic culture varies considerably across departments, so too does the amount of professional development received by graduate students prior to their first academic position. Rather than being satisfied with the current "haves" versus "have-nots" situation that characterizes professional development in the modern academic landscape, we hope that this book and the related literature serves to promote dialogue and change within departments about how early career faculty can best be supported on their paths toward tenure.

Each of us can act to improve the culture and climate of our departments, institutions, and disciplines. In large, complex institutions such as universities, it is too easy to hold others responsible for problems when, in fact, individual and collective action can be effective in implementing change. Graduate students and early career faculty should see themselves as potential contributors to the process of change. We are offering this book to foster a view of the early career professor not as someone who works solely to survive, but as someone who can promote change and participate in the construction of healthy, nurturing professional environments on campus. The benefits of this approach to empowerment will be felt by those in the early career stage, but senior faculty and administrators will also enjoy the opportunity to strengthen the overall quality of academic work in their department.

We believe that such change will involve questioning our assumptions about academic culture and climate. As Bauder (2006) has written, the process of academic socialization has been largely left unquestioned. Fundamental distinctions such as those between teaching, research, and service need to be interrogated in the same way that hooks (1994), Freire (1970), and Freire and Faundez (1989) have questioned the hidden assumptions and power relationships of pedagogical practice. Arbitrary institutional barriers, which can often marginalize some academics by gender, sexuality, age, family status, nationality, or personal characteristics irrelevant to scholarly and scientific achievement, need to be changed. If higher education is to continue to attract the most talented intellects as well as diversify to better reflect the larger demographic and cultural trends of society, many improvements should be considered. Our hope, to start, is that the topics of this book will be more thoroughly integrated into the graduate curriculum through seminars, workshops, and informal exchanges. We also hope that this book will help to foster a culture of support for early career faculty and, as a result, serve to counterbalance the often overly competitive aspect of academic life.

Our call for action stems, however, not from a sense of crisis, but from a sense of opportunity. In the case of geography, for instance, there are two major reasons to think carefully about the future and to prepare for its challenges. As Solem and Foote (2004) note, geography graduate students are starting degree programs at a time when the discipline has never been stronger. Undergraduate and graduate student enrollments are at historic highs, new departments continue to appear on the national map, and graduates enjoy an ample choice of employment opportunities in public and private sectors (Murphy 2007).

Accompanying this growth, however, are numerous challenges: the racial and ethnic diversity of the discipline remains very low, geography is still absent in many of the nation's elite universities, and the discipline is still often misunderstood by large segments of the American population having little or no geography preparation in school or higher education. New geography professors will also be affected by larger trends affecting higher education such as changing student demographics, new classroom technologies and course delivery systems, increasing reliance on part-time and adjunct instructors, shifting tenure policies, and pressures to hold higher education institutions more accountable for the quality of teaching and learning. The need, then, is to be ready to respond proactively to these trends and to the changing world of American higher education.

USING THIS BOOK

Aspiring Academics is designed to accommodate the diverse ways of offering support to faculty. First, the book can be used as a self-mentoring guide by graduate students and early career faculty. It can be read all at once, or chapter by chapter as needed. One approach would be to use the book as a reference for gaining a disciplinary perspective on, for example, classroom issues that are covered by a campus center for teaching and learning. This would help reinforce the idea, stated earlier, of how research on higher education can enhance disciplinary practices.

Second and equally important, the book is intended to be used in graduate seminars, in two different ways. In some cases, this book may work as a primary or supplemental text in some existing seminars such as those on theory and methods, history and philosophy, research and publishing, or on issues of learning and teaching. Additionally, we hope that this book will be an encouragement for the creation of new seminars on professional development in many graduate programs. Perhaps a dozen geography programs already have seminars of this sort offered for one to three credits designed for graduate students near the end of their studies and close to their first appointment.

Finally, the book can be used for workshops, reading groups, brown-bag sessions, and in other informal settings. Such sessions promote sharing of ideas, strategies, problems, and solutions, an important part of any professional development effort. The chapter activities on the book's web site include sample procedures that workshop facilitators can use to introduce topics with graduate students and faculty in a department. We also think these will be particularly helpful for senior faculty who might be interested in introducing these chapter topics in seminars and would benefit from some suggestions for strategies for promoting discussion.

The book is divided into three sections. The first, focusing on career planning and personal management issues, takes up what we view as key issues spanning all areas of academic achievement. The chapters touch on the way we manage time, plan our careers, cultivate collegial relationships, balance professional and personal lives, and prepare for tenure. There are the "big picture" issues we see as fundamental to success. These issues interlock and we have tried to draw out interconnections among these topics as well as to the other chapters in the book.

The second addresses issues revolving around our roles as teachers and advisors. These roles often produce the greatest stress and anxiety because, as was mentioned earlier, they are not usually addressed in most graduate programs. Here, in *Aspiring Academics*, we have picked out five of what we see as the most important topics relating to teaching: designing significant learning experiences, strategies for active learning, student advising, ethical issues in teaching, and teaching diverse students and teaching for inclusion. From the many Geography Faculty Development Alliance (GFDA) workshops we have led, these topics seem to be of critical importance for most graduate students and early career faculty. We have saved a full range of other topics for *Teaching College Geography* (Solem and Foote 2009), including a complete guide to getting started in the college classroom. *Teaching College Geography* also includes additional chapters and materials on promoting the scholarship of teaching and learning, GIS and mapping tools for reasoning and critical thinking across the curriculum, looking beyond the lecture and developing significant learning in large classes, teaching in the field, and geography and global learning.

The final section focuses on enhancing research and writing skills, including chapters on writing competitive grant proposals, understanding ethical issues in research, preparing research manuscripts for publication, and doing interdisciplinary research. Of course, such topics are addressed in many graduate programs. But we have found, after offering numerous workshops, that many graduate students and early career faculty still have an array of questions revolving around practical issues of writing and publishing and finding grant funding. How to respond effectively to reviewer and editorial comments, how to structure grant budgets, and how to cultivate productive interdisciplinary research may seem like small questions, but they can stand in the way of successful writing and research careers.

They are also the questions new hires hesitate to ask of colleagues for fear of appearing not to know their new jobs. Of course, in these areas of research and writing there is a wealth of other resources on these topics, and the authors have cited a range of materials that will prove useful.

ACKNOWLEDGMENTS

This book is the result of the contributions of many people over several years. Since 2002 the National Science Foundation has funded a project to examine academic professionalization in geography and provide early career faculty with the theoretical and practical knowledge needed to succeed in their careers of research, teaching, and service. That project, the Geography Faculty Development Alliance (GFDA) (DUE 0089434), is built around a program of summer workshops as well as follow-up seminars, panel discussions, and paper sessions held at professional meetings of the Association of American Geographers (AAG) and the National Council for Geographic Education (NCGE).

A second major project led by the AAG and known as Enhancing Departments and Graduate Education (EDGE) in Geography (NSF ROLE 0439914) not only complements, but considerably expands and extends the GFDA objectives by including research on individuals at earlier stages of professional development, while also focusing on the challenges of preparing geographers for professional careers in government, business, industry, and the nonprofit sector. Together, the GFDA and EDGE projects provide this book with a disciplinary foundation of empirical data, practical resources, and most importantly of all a national cadre of graduate students and early career faculty who have volunteered their time and energy toward helping us understand the needs of the new generation of academic geographers, and how their future growth can be supported by individuals, departments, institutions, and professional associations. The editors would like to thank all of the participants and contributors to both projects. Many of the authors participated in GFDA workshops, and the materials for many chapters in this book were first tried there.

This book has been a pleasure to edit, due in large part to the generosity of the many authors and reviewers. When we first began detailed planning two years ago we were overwhelmed by our call for contributors. There seems to be a widespread recognition that more and better professional development is vital to geography and all neighboring disciplines—that the time has come to reshape and reform the discipline in ways which will maintain its vitality long into the future.

We wish to thank in particular the advisors, research assistants, and volunteers who served for multiple years on the EDGE and GFDA projects: Rachelle Brooks, Ivan Cheung, Tim Conroy, Teresa Dawson, Dee Fink, Robin Friedman, Gayathri Gopiram, J. W. Harrington, Jr., Matt Koeppe, Vicky Lawson, Jenny Lee, Jongwon Lee, Duane Nellis, Mark Purdy, Doug Richardson,

Megan Overby, Sue Roberts, Fred Shelley, Beth Schlemper, Patricia Solis, Adam Thocher, and Antoinette WinklerPrins. Myles Boylan at the National Science Foundation has provided valuable advice on many occasions on these and other projects.

We also acknowledge the important service of the reviewers who provided detailed, constructive feedback on earlier drafts of the chapters in this book: Derek Bruff, Bill Buskist, Cindy Finelli, Susan Gallagher, Trav Johnson, Lisa Kornetsky, Karron Lewis, Tom Luxon, Yonna McShane, Mindy McWilliams, Judy Miller, Gail Rathbun, Richard Reddy, Kay Sagmiller, and Lynn Sorenson.

References

Bauder, H. 2006. Learning to become a geographer: Reproduction and transformation in academia. *Antipode* 38 (4): 671–79.

Becher, T. 1989. *Academic tribes and territories: Intellectual enquiry and the cultures of disciplines*. Buckingham, U.K.: Open University Press.

Boice, R. 1992. *The new faculty member: Supporting and fostering professional development*. San Francisco: Jossey-Bass.

———. 2000. *Advice for new faculty members*. Needham Heights, MA: Allyn & Bacon.

Boyer, E. L. 1990. *Scholarship reconsidered: Priorities of the professoriate*. Princeton, N.J.: Carnegie Foundation for the Advancement of Teaching.

Cameron, K. S. and D. R. Ettington. 1988. The conceptual foundations of organizational culture. In *Higher Education: Handbook of Theory and Research, Volume IV*, ed. John C. Smart, 356–396. New York: Agathon Press.

Council of Graduate Schools. 2003. *Inclusiveness series*. Washington, DC: Council of Graduate Schools.

Fink, L. D. 1984. *The first year of college teaching*. San Francisco: Jossey-Bass.

Freire, P. 1970. *Pedagogy of the oppressed*. Trans. M. B. Ramos. New York: Herder and Herder.

Freire, P., and A. Faundez. 1989. *Learning to question: A pedagogy of liberation*. Trans. T. Coates. New York: Continuum.

Gaff, J. G., A. S. Pruitt-Logan, and R. A. Weibl. 2000. *Building the faculty we need: Colleges and universities working together*. Washington, DC: Association of American Colleges and Universities.

Glassick, C. E., M. T. Huber, and G. I. Maeroff. 1997. *Scholarship assessed: Evaluation of the professoriate*. San Francisco: Jossey-Bass.

Golde, C. M., and T. M. Dore. 2004. The survey of doctoral education and career preparation: The importance of disciplinary contexts. In *Paths to the professoriate: Strategies for enriching the preparation of future faculty*, eds. D. H. Wulff and A. E. Austin, 19–45. San Francisco: Jossey-Bass.

Golde, C. M., and G. E. Walker, eds. 2006. *Envisioning the future of doctoral education: preparing stewards of the discipline*. San Francisco: Jossey-Bass.

Hardwick, S. W. 2005. Mentoring early career geography faculty: Issues and strategies. *The Professional Geographer* 57 (1): 21–27.

Healey, M. 2003. Promoting lifelong professional development in geography education: International perspectives on developing the scholarship of teaching in higher education in the twenty-first century. *The Professional Geographer* 55 (1): 1–17.

Healey, M., and A. Jenkins. 2003. Discipline-based educational development. In *The scholarship of academic development*, eds. H. Eggins and R. Macdonald, 47–57. Buckingham, U.K.: Open University Press.

hooks, b. 1994. *Teaching to transgress: Education as the practice of freedom*. New York: Routledge.

Lee, J. 2007. The shaping of the departmental culture: Measuring the relative influences of the institution and discipline. *Journal of Higher Education Policy and Management* 29 (1): 41–55.

Moss, P., A. Cravey, J. Hyndman, K. K. Hirschboeck, and M. Masucci. 1999. Towards mentoring as feminist practice: Strategies for ourselves and others. *Journal of Geography in Higher Education* 23 (3): 413–27.

Murphy, A. 2007. Geography's place in higher education in the United States. *Journal of Geography in Higher Education* 31 (1): 121–141.

Petersen, M. and M. Spencer. 1990. Understanding academic culture and climate. In *Assessing academic climates and cultures*. ed. W. G. Tierney. San Francisco: Jossey-Bass.

Rosser, V. 2004. Faculty members' intentions to leave: A national study on their work-life and satisfaction. *Research in Higher Education* 45 (3): 285–309.

Solem, M. and K. Foote, eds. 2009. *Teaching College Geography: A Practical Guide for Graduate Students and Early Career Faculty*. Upper Saddle River, NJ: Pearson Education, Inc.

Solem, M. N., and K. E. Foote. 2004. Concerns, attitudes, and abilities of early-career geography faculty. *Annals of the Association of American Geographers* 94 (4): 889–912.

Sorcinelli, M., and A. Austin, eds. 1992. *Developing new and junior faculty*. San Francisco: Jossey-Bass.

Wulff, D. H., and A. E. Austin, eds. 2004. *Paths to the professoriate: Strategies for enriching the preparation of future faculty*. San Francisco: Jossey-Bass.

ABOUT THE EDITORS

Michael Solem is Educational Affairs Director for the Association of American Geographers, where he directs the Enhancing Departments and Graduate Education in Geography (EDGE) project and the Center for Global Geography Education initiative funded by NSF. Dr. Solem is the external evaluator for the Geography Faculty Development Alliance and the Graduate Ethics Education for Future Geospatial Technology Professionals projects. He currently serves as the North American coordinator of the International Network for Learning and Teaching Geography in Higher Education (INLT), is associate director of the Grosvenor Center for Geographic Education at Texas State University-San Marcos, and leads the AAG's efforts with the Carnegie Academy for the Scholarship of Teaching and Learning program. He received the *Journal of Geography in Higher Education's* biennial award for promoting excellence in teaching and learning for his research with Ken Foote on faculty development in postsecondary geography.

Kenneth Foote is professor of geography and former department chair at the University of Colorado at Boulder. His research spans cultural and historical geography, GIScience and cartography, and geography education, especially the application of instructional technologies and issues of geography in higher education. Ken has led the NSF-funded Geography Faculty Development Alliance since 2002 and is co-principal investigator of the Enhancing Departments and Graduate Education in Geography project with Michael Solem and Janice Monk. He has served as president of the National Council for Geographic Education (2006) and national councilor of the Association of American Geographers (2002–2005). Ken received the Association of American Geographers' 1998 J. B. Jackson Prize for his book *Shadowed Ground: America's Landscapes of Violence and Tragedy* and the association's 2005 Gilbert Grosvenor Honors in Geographic Education.

Janice Monk is professor of geography and regional development at the University of Arizona where she currently teaches a graduate course in professional development. Dr. Monk is also Research Social Scientists Emerita having served for over twenty years as director of the University's Southwest Institute for Research on Women. There she was involved with interdisciplinary research, faculty development, and community outreach projects in the region and in collaboration with colleagues in Mexico. Jan's interests and publications span feminist geography, international collaboration, history of women in geography, and faculty development in higher education. She is co-principal investigator of the Association of American Geographers' project "Enhancing Departments and Graduate Education in Geography," participated for four years as a staff member of the Geography Faculty Development Alliance, and also served the AAG as president in 2001–2002.

ABOUT THE AUTHORS

Stanley D. Brunn is professor at the University of Kentucky with multiple interests in human/environmental geography. He served as editor of *The Professional Geographer* (1982–1987) and *Annals, AAG* (1988–1993) and has written and edited more than 100 articles and chapters and more than 20 books on political, social, and urban geography. Many of these projects were with former graduate students and colleagues in other countries. He has taught at the University of Florida and Michigan State University and for short periods in many European and Central Asian universities. He has received AAG Honors and a Distinguished Mentor award from the National Council for Geographic Education. His most recent edited volumes address diverse subjects such as the geopolitics of terror, Wal-Mart, gated communities, and music geography. His current interests include the impacts of mega-engineering projects, state-identity issues, and cybermapping. In Fall 2007 he was Fulbright Scholar in Semey, Kazakhstan.

So-Min Cheong is an assistant professor in the Department of Geography at the University of Kansas. Her research interests are the human dimensions of coastal change, community-based analysis of environmental and economic changes, and integration of social and natural sciences in human–environment research. She has published a number of articles on human–environment interactions, coastal management, fisheries, and tourism, and is preparing a book on qualitative methods for place-based research. Dr. Cheong is currently a visiting professor at Stanford University working on interdisciplinary coastal and environmental issues in the Yellow Sea and the Pacific U.S.

Dee Fink is a nationally recognized expert on college teaching and faculty development. After receiving his doctorate from the University of Chicago in 1976, he accepted a faculty position at the University of Oklahoma. In 1979 he founded the Instructional Development Program at the University of Oklahoma and served as its director until his retirement from Oklahoma in May 2005. He was president of the POD Network (Professional and Organizational Development) in Higher Education (2004–2005), the primary professional organization for faculty developers. Currently Dee works as a national consultant in higher education.

Eric J. Fournier is associate professor and chair of the Geography Department at Samford University in Birmingham, Alabama. While trained as an urban geographer, most of his recent work involves the study of teaching and learning in undergraduate geography with a particular interest in active and problem-based learning. His current research efforts involve spatial thinking and the infusion of technology in multidisciplinary, general education courses. He was co-principal investigator for the NSF-funded AEGIS (Academic Excellence through Geographic Information Science) project that helped train faculty to use spatial technologies and perspectives in their introductory classes. He is also interested in issues related to faculty development and training and has been a workshop leader for the Geography Faculty Development Alliance.

Melissa Ganus is a social entrepreneur (focused on outcomes beyond profits), researcher, and curriculum design consultant. Melissa's experiences are wide ranging, working at Microsoft in the 1980s, running two Seattle-based nonprofits in the 1990s, and teaching her first college courses in 1996. As a first-time teacher, she discovered her academic experiences had not taught her how to create powerful and relevant learning experiences for others. Since then, she has developed an expertise in student-centered adult learning, curriculum design, and applied psychology (positive, educational, and social). When starting on her doctorate in educational leadership, she became acquainted with Dee Fink's work and was subsequently able to assist him on a few of his projects, including coauthoring the chapter in this book. In addition to her studies, she provides consulting, curriculum design, workshops, and related resources to help educators make their course planning and presentations easier and more effective.

Iain Hay is a human geographer with particular interests in geographies of oppression. He is professor of geography at Flinders University, South Australia, where he recently served for six years as Head of the School of Geography, Population and Environmental Management. Dr Hay's eight books include *Research Ethics for Social Scientists: Between Ethical Conduct and Regulatory Compliance* (with Mark Israel). He has received research awards from organizations such as the British Academy, the Fulbright Commission, and the Australian Institute of Urban Studies. Professor Hay is currently on the editorial boards of ten journals including *Applied Geography, Geographical Research*, and *International Research in Geographical and Environmental Education*. He was among the founding editors of *Ethics, Place and Environment* and was the first Australasian commissioning editor for *Journal of Geography in Higher Education*. He is currently establishing a new journal entitled *Cultural Landscapes*. In 2006, Professor Hay received the Prime Minister's Award for Australian University Teacher of the Year.

Mark Israel has a degree in law and postgraduate qualifications in sociology, criminology, and education. His most recent books include *Research Ethics for Social Scientists: Between Ethical Conduct and Regulatory Compliance* (Sage, 2006 with Iain Hay) and *Crime and Justice: A Guide to Criminology* (Lawbook Company, 2006). Mark is a member of the subcommittees responsible for research ethics of both the British Society of Criminology and the Australian and New Zealand Society of Criminology, is associate editor of *Criminal Justice Ethics,* and is Pacific Rim Editor for *Critical*

Criminology: an International Journal. He went to Flinders University in Australia in 1993 and now teaches a range of topics in criminology in the School of Law where he is also faculty associate head (Research). Mark won the Prime Minister's Award for Australian University Teacher of the Year in 2004, the Radzinowicz Memorial Prize from the *British Journal of Criminology* in 2005, and the Critical Criminologist of the Year Award from the Critical Criminology Division of the American Society of Criminology in 2006.

Christine L. Jocoy is a teacher/scholar and assistant professor of geography at California State University, Long Beach. Her research and teaching interests include economic and urban geography, regional restructuring, globalization, learning theory in corporate decision-making, urban social policy, and pedagogy and best practice in geography education. Common to her work is the quest for understanding how students, individuals, or organizations learn, how decision makers use relevant information, and why managers often ignore important data. She recently received a grant from the METRANS Transportation Center to study urban social and transportation policy related to homelessness in Long Beach, California.

James Ketchum is special projects coordinator and newsletter editor for the Association of American Geographers in Washington, DC. He earned his Ph.D. in cultural geography from the Maxwell School of Citizenship and Public Affairs at Syracuse University, under the direction of Don Mitchell. After graduating with a master's degree in geography from the University of Illinois, Urbana-Champaign, James worked at art museums in Cincinnati and Virginia Beach before pursuing a Ph.D. He has taught at Clarion University in Pennsylvania, Syracuse University, Tidewater and Parkland community colleges, and most recently at Keuka College in Keuka Park, New York. He is currently co-editing a book, surveying recent developments in geography and the humanities, to be published in 2009.

Minelle Mahtani is an assistant professor in the Department of Geography and Planning, and the Program in Journalism, at the University of Toronto. She was previously a professor at Lang College at the New School of Social Research in Media Studies and held the Killam postdoctoral fellowship in the Department of Geography and the School of Journalism at University of British Columbia. Minelle is completing a book with UBC Press on the experiences of mixed race women in Toronto. Her research explores the intersection between the diversity initiatives in international media networks and the negotiation of space, power, gender, and identity in these sites.

M. Duane Nellis is provost and senior vice president of Kansas State University. As K-State's Provost he has focused on an institutional priority setting process that recognizes the need for a more engaged university including forward looking faculty profiles, greater internationalization, increased levels of entrepreneurship, and enhanced learning experiences for students. Dr. Nellis is also professor of geography with specialization in remote sensing and geographic information systems as they apply to natural resource systems. As a geographer, his research has been funded by more than fifty grants from such sources as NASA, the National Science Foundation, National Geographic Society, the U.S. Agency for International Development, and the U.S. Department of Agriculture. Duane has served in various professional leadership positions nationally, including most recently as past president of the Association of American Geographers (2002–2003) and as past president of the National Council for Geographic Education.

Adrienne M. Proffer is an academic advisor in the Center of Undergraduate Academic Advisement at the University of Central Oklahoma. She advises students who are seeking bachelor of arts degrees in geography, as well as students who are seeking other degrees offered by the College of Liberal Arts. She holds an M.A. in geography from the University of Oklahoma, where she served as Fred Shelley's research assistant, and a B.A. in geography from the University of Texas at San Antonio.

Susan M. Roberts did her undergraduate studies at the University of Leicester (U.K.) and her graduate work at Syracuse University. She has been on the faculty in the Department of Geography at the University of Kentucky since 1991, where she is also a member of the Committee on Social Theory and the Gender and Women's Studies Program. Sue's research interests lie in political economy, critical social theory, and feminism, and she has received research support from the National Science Foundation, among others. She has co-edited books, and authored book chapters and articles in a range of journals. Sue has worked with many graduate students and contributed to the preparing future faculty program at the University of Kentucky. She was very honored to be awarded the University of Kentucky's Sturgill Prize in 2006 in recognition of her contributions to graduate education.

Beth Schlemper is a visiting professor at the University of Toledo in the Department of Geography and Planning. She is also a principal researcher for the AAG's "Enhancing Departments and Graduate Education (EDGE) in Geography" project. Prior to arriving in Ohio in 2005, Dr. Schlemper was an assistant professor in the Department of Geography–Geology at Illinois State University for four years. Her primary research interests have been related to the construction of regional identity and scale. Recent publications have incorporated methods from a variety of disciplines, such as geography, history, sociology, and anthropology. Dr. Schlemper has published scholarly articles in the *Journal of Cultural Geography* and the *Journal of Historical Geography*. She also has a chapter "The Borders of the Holyland of East-Central Wisconsin" in an edited volume entitled *Wisconsin Germans: Land and Life* (University of Wisconsin Press 2006).

Patricia Solís is director of Research and Outreach for the Association of American Geographers where she designs and implements special projects to promote studies, education, and public understanding of geography. She specializes in sustainable development, water resource issues, geographic technologies, and international–intercultural affairs, particularly with respect to Latin America. Dr. Solís has experience writing at least fifty major grant proposals, has won more than $4.7 million in external grants, and has administered or evaluated projects in excess of $6 million. Her list of awards includes corporate, foundation, government sources, fellowships, and international donors, such as the Ford Foundation, state block grant programs, the U.S. Department of Transportation, U.S. Department of Agriculture, the InterAmerican Development Bank, the Kellogg Foundation, the Fulbright Program, the U.S. National Science Foundation, and others. She is a Certified Grants Specialist by the National Grant Writers Association and is a member of the International Association of Grant Writers and Nonprofit Consultants.

Fred M. Shelley is professor of geography and chair of the Department of Geography at the University of Oklahoma. He holds a Ph.D. in geography from the University of Iowa and has taught at the University of Southern California, Florida State University, and Texas State University—San Marcos. Fred's research and teaching interests include the political, economic, and cultural geography of North America, and he has published over 100 articles and books on these and related subjects. He is a past member of the Council of the Association of American Geographers (AAG).

Antoinette M.G.A. WinklerPrins is an associate professor and graduate program coordinator in the Department of Geography at Michigan State University. She has conducted research on smallholder livelihoods, primarily in the Brazilian Amazon, for fifteen years. She has been funded by NSF, the American Philosophical Society, and Michigan State University, and her research has been published in numerous journals and edited volumes including *The Geographical Journal*, *Latin American Research Review*, *Agriculture and Human Values*, and *The Journal of Latin American Geography*. Antoinette has won several teaching awards, including a Teacher-Scholar Award and a Lilly teaching fellowship. Her innovative development of an award winning online version of World Regional Geography has helped propel her department to the forefront online geographical education. As graduate program coordinator she oversees an increasingly large and diverse graduate program in one of the largest departments of geography in U.S. She is actively involved in providing early career professional development in her department by conducting workshops for senior graduate students and mentoring junior faculty.

Craig ZumBrunnen is professor of geography and member of the Jackson School of International Studies at the University of Washington (UW). His professional training is an interdisciplinary synthesis of diverse natural and social sciences. Since 1968, he has primarily focused on interdisciplinary research and field experience in the former Soviet Union dealing with urban, natural resource management, energy, and environmental pollution problems. He has been involved in interdisciplinary environmental teaching for nearly thirty years and since 1989 has made increasing use of experiential learning and small group processes in his Geography classes, including his classes on mathematical modeling of human–economic–ecological interactions. In 1999, he was selected to the first cohort of the UW's Institute of Teaching Excellence and in 2000 received the UW Department of Geography's Award for Excellence in Undergraduate Teaching. Recently, he has integrated service learning and interdisciplinary experiential active team-learning processes into several courses, including his efforts as an active member of the UW's multiyear, team-taught course sequences in urban ecology.

Career Planning and Personal Management

My number one issue is to learn to do more in less time and work more effectively.

I have just recently become a dad. I have a young boy and so I definitely feel the need to be at home. But I want to be able to provide for him by being successful.

I could identify three different senior faculty members who would give me very different pictures about what is required for tenure.

And long term . . . I think there is a lot to be said about a kind of active sharing of ideas. . . .

Solem and Foote 2004, 894

These comments by early career faculty address some of their key challenges, hopes, and desires for advice. Perhaps, as a graduate student or new professor, they embody some of your own concerns and aspirations. Although graduate education provides rigorous preparation for research and often some experience (though perhaps not systematic preparation) in teaching, it rarely systematically addresses managing the competing demands of a new job in a new place. The graduate years have their own stresses—being short on money, coping with too-long reading lists, meeting end-of-semester deadlines for multiple papers, being thrust into teaching courses for which you may be ill-prepared, getting past a writing block to finish the dissertation, and others—but throughout, you probably have been working with a peer group that generally offers social and intellectual support. Now you have reached a new peak—doctorate in hand, you have landed a faculty position. Somehow, it seems not at all like graduate school. You are new in a community where others seem settled, you are faced with several teaching preparations and with expectations that you will publish, serve on departmental committees, and work toward tenure (or, if in a temporary appointment, keep yourself active on the market). Where and how do you find support?

We have chosen to open this book with "active sharing of ideas" to assist aspiring academics in meeting key challenges they have identified. Time management looms large and so we address it in Chapter 1. So too do questions of how to bring together personal aspirations and the demands and opportunities of the new context. What should you aim to accomplish now and over the next few years? What resources do you need to move ahead? Chapter 2 takes up these issues by focusing on career planning through the lens of linking personal and professional aspirations. Chapter 3 extends discussion of concerns that are threaded through the previous two by focusing on collegiality. What should you expect of colleagues, and what can you contribute to your department, institution, and profession? Beyond the job, how do you have a life that is satisfying while being responsible to your institution, partner, family, and friends? Those questions form the basis for Chapter 4. Then we turn in Chapter 5 to the preoccupation that weighs on early career faculty: what will it take to jump over the tenure hurdle? What advice should we seek and from whom?

Our emphasis in this section is somewhat different from other guides for early career faculty, which often begin with issues of teaching and research and touch on these other professional issues, if at all, almost as an afterthought. Our belief is that the issues addressed in this section, though rarely brought up in graduate curricula, are actually among the most fundamental to success. By drawing on the literature of higher education and the resources of experienced academics and of peers, early career academics can reduce stresses and thrive. It is in that spirit that we offer the following chapters and related exercises.

OVERVIEW OF THE CHAPTERS

In his chapter on time management, Ken Foote makes two key points: the importance of sorting out priorities according to their relative importance and urgency and of developing strategies of moderation and balance in approaching tasks. Much can be accomplished by working in small but regular blocks of time; by learning where, when, and how to use time most effectively; by seeing work as a social process that engages the support of others; by taking breaks; and by learning when to say "no." He recommends that you identify a couple of strategies to practice in the short and long terms. In related exercises he provides suggestions of strategies that early career academics have found helpful and sets up an approach to evaluating your own use of time.

Janice Monk, a senior academic, and Christine Jocoy, in the early years of her career, come together to reflect on career planning. They argue that it is important to have a "road map" but to be prepared for changes along the journey. Important first steps are to clarify your personal and professional values, and to identify your strengths and those areas in which you would like to improve. You can then proceed to specify what activities, resources, and supports you will need to work toward your personal professional goals in the different areas of work, to think how tasks can be integrated, and to set time lines for attaining them. In creating, presenting, and reflecting on her own development plan Christine offers a realistic model.

The chapter by Duane Nellis and Susan Roberts takes up the topic of collegial relations—often a vaguely understood concept. Their discussion is set within the context of contemporary changes in higher education and attends to local settings in the department and on campus as well as to wider professional arenas, for example, in professional associations. Above all, they highlight the importance of focusing on constructive relationships, though not ignoring that at times there may be difficult situations that call for respectful dissent. Among principles they advocate are being a good listener and being proactive and strategic in ways that will reduce stress and advance personal goals while also supporting collective endeavors. Their related activities provide resources for reflection on the meanings and place of collegiality and also offer case studies that require analysis of situations that call for creative responses.

As we have suggested, the personal and the professional lives are intimately intertwined. We do not work or live as isolated beings, despite clichés of the academic as one who lives in an "ivory tower." Increasingly, administrators and faculty (individually and collectively) are working to create policies and practices that will accommodate personal life, especially as institutions have become more diversified and more women have joined the professoriate. Beth Schlemper and Antoinette WinklerPrins open their chapter by noting that the academic life, though requiring many hours of work, offers satisfactions and flexibility not found in other professions. Complementing Ken Foote's chapter on time management, they attend to multiple challenges and opportunities for balancing personal and professional life, especially in relation to children and family roles. They discuss the timing of life decisions, sorting out priorities when searching for positions. They consider diverse arenas: departmental management and relations with colleagues, teaching, field work, and conference attendance. They point out the ways in which the demands of positions vary across types of institutions; indicate how "family friendly" policies are being implemented; and list resources that describe supports such as family leave, day care, and partner placement. Schlemper and WinklerPrins acknowledge that needs vary at different points in life and according to personal circumstances. Like the authors of other chapters, they recognize that collegiality and cooperation are important parts of making life work for the individual faculty member and for shaping supportive climates within institutions.

The final chapter in this section, by Susan Roberts, takes up the too often mysterious process of tenure—what it means, how to attain it, and what lies beyond it. She explains the concept as not simply the means to job security, but as a privilege that offers protections for doing creative and original work, the highlights of an academic career. Since expectations for tenure vary considerably between types of institutions, especially between those where doctoral degrees are awarded and others, she urges aspiring academics to seek out information early about institutional rules and procedures (e.g., how will joint appointments be handled?). She reviews ways in which research, teaching, and service are integrated and evaluated. She reminds us of the values of seeking support from (and giving it to) mentors and peers. Such support may be especially critical for "non-traditional" faculty: among them are women and members of sexual, ethnic, and racial minority groups on whom special demands and expectations are often placed. Her "bottom line" is that tenure should be fulfilling personally and is a professional validation, and that there are many measures we can take to prepare for and succeed in the review process.

Together, these chapters complement each other and introduce important aspects of the academic life that are rarely discussed systematically and often not openly. We hope these chapters will be useful and will stimulate you to engage in conversations with peers and mentors, to be proactive in planning your career.

References

Solem, M., and K. Foote. 2004. Concerns, attitudes, and abilities of early-career geography faculty. *Annals of the Association of American Geographers* 94(4): 889–912.

CHAPTER 1

Time Management

Kenneth E. Foote

Time management is perhaps *the* major source of stress and anxiety for graduate students and early career faculty. This has a been a consistent finding over many studies (Fink 1984, 1988; Seldin 1987; Boice 1992; Sorcinelli 1992; Gmelch 1993; Fisher 1994; Reis 1997; Edworthy 2000) and is expressed often in quotes like "My number one issue is to learn how to do more in less time and work more effectively because I literally work all of the time. My health has really suffered because of it" and "One of the biggest challenges is juggling teaching, research, and service" (Solem and Foote 2004, 894). Part of the reason for this stress is that time management underlies many of the topics covered in this book. Balancing the time demands of competing tasks is fundamental to preparing for teaching, gaining the most from effort spent on research and writing, getting ready for promotion reviews, coping with responsibilities outside of work, and even building effective relationships with colleagues and peers. Rather than addressing each of these issues separately, this chapter will review current research on time management with the aim of drawing out common themes and strategies that can be applied as needed to the many things we do as scholars and scientists.

As you read this chapter, try to put its ideas to use by identifying two time management strategies you can begin to use immediately in your work. Since some strategies take longer to implement than others, pick one you can put into practice over the next semester and one you can apply to your work over the next year. Two activities accompany this chapter that will help you identify suitable strategies from the many discussed. Activity 1.1 is designed to encourage the sharing of time-management strategies and tips among peers, friends, and colleagues. Some research about time management indicates that people often feel they have to "go it alone" in solving their problems when, in fact, friends, family, and colleagues can offer considerable help. Activity 1.1 emphasizes time management as a social process and is designed to encourage such discussion and sharing. Activity 1.2 involves keeping a time log. Time logs are recommended in most self-help guides because they assist in pinpointing exactly where useful changes can be made in busy schedules.

WHAT DO WE KNOW?

One of the problems in finding help with time management is that relevant research is aimed at the world of business, not higher education. The result is that while there are many self-help guides providing good advice on office routine found in the business sections of most bookstores (Hindle 1999; Morgenstern 2000; Davidson 2001), these books address only a portion of what academics do. The difference stems from the fact that academic work usually involves juggling a greater number and greater variety of projects that demand more varied skills than those required of many other professions. For example, teaching is a highly public activity that can require good presentational, leadership, and interpersonal skills and offers relatively immediate feedback and rewards. Research, on the other hand, tends to be a relatively private activity involving self-motivation, self-discipline, and concentration, and is an activity with relatively distant payoffs in terms of rewards and recognition. Service activities entail these and other talents but offer relatively few immediate and tangible rewards. Few professions require such a mix of talents on a daily basis. The need to do well in so many domains results in the extra pressures of academic life. The promotion and tenure process also adds pressure. Once beyond the dissertation, rarely will a single article, book, grant, teaching award, or other accomplishment tip the balance unequivocally for promotion. The key to success for most faculty is usually found in doing well in a variety of projects which, taken together, express their intellectual range, scholarly trajectory, teaching abilities, and collaborative skills in working with students and colleagues.

This means that time management is not simply a process of making a work list and ranking the items in priority order. Such a list would imply that teaching, research, and service tasks are commensurate and can be readily compared and linearly ranked when, in fact, they are usually very difficult to judge side by side. A better way of looking at this problem can be found in Covey's influential *The 7 Habits of Highly Effective People* (1989). In the chapter "Habit 3: Put First Things First," Covey classifies work tasks within a two-dimensional matrix (rather than as a one-dimensional list) by arranging tasks by their importance and urgency. In Figure 1.1, I have applied Covey's distinctions between *Important–Unimportant* and *Urgent–Not urgent* to tasks typical of academic work. Covey's point would be that some of the most important work we do lies in the *Important/Not urgent* (upper right) quadrant of the matrix, but the tasks in the urgent column (upper and lower left) can too often draw our attention away from more important work. And, if we

FIGURE 1.1 A view of academic work derived from Stephen Covey's Important–Unimportant and Urgent–Not urgent matrix of work priorities.

	Urgent	Not Urgent
Important	Meeting responsibilities of current term—class schedules, grading, committee work. Responding to deadline-driven projects such as submission of abstracts, grants, and manuscripts. Responding to pressing problems of students and colleagues.	Networking. Many writing and research projects. Seeking funding for teaching and research projects. Reflecting and improving upon teaching and curriculum. Mentoring and helping others.
Non Important	Interruptions. Some calls. Some mail and e-mail. Some reports. Some meetings.	Some calls. Some mail and e-mail. Some reports. Some meetings—both intramural and exramural.

Source: Adapted from Covey 1989, 151.

spend all of our time addressing work tasks in the upper-left quadrant, we can quickly become exhausted without ever getting to some of our important, but not urgent work. Covey's point is that it is essential for people to organize their activities around priorities but most critically, individuals can take charge of how their work is arranged in this matrix. By prioritizing tasks in the *Important* row (upper left and right), we can recognize unimportant tasks for what they are and reduce their claims on our time. Also, by focusing on taking charge of our schedules and planning ahead for deadlines, we can relegate more tasks to the *Important/Not urgent* (upper right) quadrant.

Covey provides an appealing vision of effective time management, but his writings are neither aimed expressly at academics, nor based on empirical research. To find such work, one must turn to Boice and the body of research he has published based on surveys, observational study, intervention programs, and workshops (Boice 1991, 1992, 1997, 2000). Boice's major recommendation can be captured in one word—moderation—but his findings offer specific suggestions for enhancing time management and reducing stress.

Boice has found that, in part, reducing stress involves facing several misperceptions about time management. The most common is that we are always "busy." Indeed, "busyness" is the single most common excuse for low productivity but, in fact, busy or not, most people are not very good at remembering exactly how they spend their time. When observed, most individuals have blocks of time available during the day—sometimes only 15–30 minutes at a time, sometimes longer, but they *perceive* these periods to be too short to be valuable.

This misperception is compounded by the view that large blocks of time are required for the most important projects. But Boice would maintain that much productive work can be accomplished in small blocks—both on the "big" projects and on shorter tasks like writing references and answering e-mail that clear time for longer periods of concentration. One of the main objections given by many academics to working on "big" projects in small blocks is that they cannot build the momentum needed to work or write effectively. Boice (1990, 1997) has found, however, that most people working in shorter, more regular periods actually sustain their momentum from session to session without additional warm-up.

Boice's work implies that improvement in time management often involves confronting misperceptions, but he also makes the point that improvement can also entail changing habits. One of the most prevalent Boice calls "bingeing," or waiting too long to start a project, working to exhaustion to complete it, then turning to the next project late, working to exhaustion to meet its deadline, and then turning yet again to another project which can be finished only through further phenomenal effort. Exhaustion is the end result of such bingeing. Certainly, bingeing can help us meet some deadlines but, in the long term, Boice offers evidence that steady, regular effort yields more and better work without as much stress or anxiety. Time bingeing is quite common among academics because we are often encouraged to binge in college, graduate school, and professional life—and to take pride in, even to brag about the long and exhausting hours we keep. Time bingeing is sometimes hard for people to give up because they have achieved such success using it to write theses, dissertations, articles, and proposals. But, again, Boice argues that—in the long run—a moderated approach yields higher gains.

The good news is that research by Boice and others (Ferrari, Johnson, and McCown 1995; Schouwenburg et al. 2004) indicates that time management habits like bingeing are malleable and can be changed. However, the techniques vary from person to person and can involve both short- and long-term strategies. But this brings me to an important point: When I speak of breaking habits and making changes, I am suggesting a range of strategies rather than one-size-fits-all solutions. Please consider the ideas presented below in light of your life and professional goals. For some people, a small change can make a big difference; others find that improvements involve considerable effort. For this reason, when I hold workshops, I ask participants to pick only two strategies they wish to try initially—one that they can implement over the next semester, and one that may take more effort but can be implemented over the next year. I see the strategies as falling into three general categories relating to time, place, and people.

RETHINKING TIME

Boice's main recommendation—moderation—means that working in shorter, regular periods tends to be more productive than time spent in longer and irregularly scheduled blocks. These shorter periods of concentrated thought and work have two key advantages: (1) they are easier to fit into day-to-day schedules and (2) they allow momentum to be sustained from one work session to the next. The best times for these shorter periods vary from person to person. Some people find them in the breaks between existing commitments—classes, meetings, or other regular tasks. Others set aside periods when they will have the most energy or be the most relaxed. For some, these times are in the morning—even very early—while others find their best times to think and work in the afternoon, evening, or late at night. To make this sort of regimen work, it is important to set bounds and stick to them. That means, first of all, ending other activities promptly and transitioning to work even when the previous task is not quite finished. But, more importantly, it means starting "big" tasks sooner, even when we may feel that we are not quite ready to begin and stopping work before exhaustion sets in. When work or writing is going well, it is sometimes hard to stop and perhaps lose a train of thought. But, in many respects, it makes more sense to stop at a point from which the flow of thoughts can be taken up again in the next session than to end at a point just before a difficult transition.

The sequencing of activities through the day is also something that can be planned to improve productivity. Many people report success using contingency management, that is, scheduling enjoyable, preferred activities as rewards for completing more difficult tasks. This might mean scheduling enjoyable activities—reading, running, or having a snack—as regular breaks in our work schedule or by scheduling daily or weekly activities so that the things we enjoy most are interspersed with those activities that take great effort. The key point about rethinking schedules is to take control of time and, as Covey would maintain, to "put first things first" rather than letting other people and events define the agenda for us.

RETHINKING PLACE AND WHERE WE WORK

A second key consideration is thinking carefully about where we work and the characteristics of this work environment. Many people like to establish a dedicated space for their most concentrated thought and work, one free of distractions but with the comforts they most enjoy. For many, this work space may be at home, but many enjoy working around other people in a library or coffee shop. Still others find it important to set a clear spatial separation between their work and home lives due to personal or family commitments and may decide to work only in their offices, and not at home. Again, as with setting schedules, it is important to set clear bounds so that the edges around the workspace do not blur. For instance, if you prefer to work in your campus office it may be good to set a clear distinction between when you wish to work undisturbed and when you are available to meet other responsibilities. Sometimes this involves setting times when your office is closed and you leave phone calls and e-mail unanswered, although it is always important to be mindful of the time pressures faced by our colleagues and friends and to be respectful of their needs.

HOW OTHER PEOPLE CAN HELP AND SUPPORT OUR WORK

One of the most important insights of research into time management is the importance of creating a social support system that helps us realize our goals. I think that, too often, there is a misperception that we have to solve all of our time-management problems ourselves in isolation. But, in fact, our family, friends, students, and colleagues can assist in many ways, and we can help others. Help can be as simple as letting our colleagues know our schedules so they know when we will have our door open or closed, and why. Or it might entail letting our students know when we are facing a difficult deadline, or why we will not be able to answer e-mail or phone calls on particular days, so that they understand when they can get advice.

But Boice's research findings go further—they suggest the value of embedding plans for time management in social networks. For instance, in his study of the habits of productive writers, Boice (1997) notes the value of

1. setting limits on lecture preparation time;
2. soliciting help and advice about both research-writing and teaching from colleagues;
3. writing during more weeks of the semester, so as to feel less stressed and "busy";
4. showing high self-esteem in willingness to share rough drafts, early ideas, and occasionally poor performances.

The second and fourth points are notable because they are "social" and involve asking for help and seeking support from colleagues.

Working with friends, students, and colleagues also involves judging when to say no. For good reason, most people are hesitant to turn down requests from advisors and senior colleagues. And, certainly, there are times when we have to accept important requests, however inconvenient. But there is nothing wrong with saying "Could I think about your request overnight and let you know in the morning," "Could I work on that project next semester, rather than right now," or, "Would it be possible for me to help in a different way?" Most advisors and colleagues will understand if you can provide some good reasons for declining or postponing a request, or are willing to propose alternative ways of helping with a project.

ADDITIONAL STRATEGIES

This last example indicates how Boice's and Covey's ideas can be used to organize our schedules around priorities. What if, as a newly hired assistant professor, your chair asks you to serve on the department's graduate committee? You know this will be an open-ended assignment because the committee meets once or twice a month and has to prepare a number of reports each year at unpredictable times. Plus, the committee always reviews admissions applications just before the start of the spring semester, a time when you'd like to be out of town collecting data at a field site. Why not preempt the chair's request by volunteering to coordinate the colloquium series? From helping organize a series while in graduate school, you know you can readily recruit speakers by e-mail and phone and, since the colloquia are always held at noon on Fridays, you schedule your undergraduate seminar to meet just before or after.

This is one of many suggestions I have heard in time management seminars I have held for graduate students and early career faculty (Table 1.1). The list highlights only a small sample of promising strategies I have heard from over 300 workshop participants. But this list, like others (Pfeffer 2002; Ailamaki and Gehrke 2003; Gray 2005), also reinforces a point I made

TABLE 1.1 Strategies for better time management suggested by participants in Geography Faculty Development Alliance Workshops, 2002–2006.

Writing and research:
- Keep several research projects going at once so that one is always starting, one is in process, and one is finishing.
- Don't schedule writing at night when fatigued; this only increases frustration.
- Schedule 15–60 minutes each day for reading and writing on research topics.
- Set aside 45 minutes each day for writing, but stop early if two paragraphs are drafted before the time is up.
- Leave gaps in manuscripts to fill in with details later instead of allowing them to interrupt flow of ideas.
- Don't submit conference abstracts for work that hasn't yet been done. If the research isn't near completion when the abstract is written, the pressure of completing it will only grow. Finishing the paper will involve dropping other projects, causing stress, and throwing off other deadlines.
- Use free writing to get started on new projects.
- Don't try to write final manuscript in first draft.
- Be less judgmental of your own writing.

TABLE 1.1 Continued

Preparing for classes:
- Spread course preparation throughout the semester rather than trying to have everything ready at the start of the semester.
- Develop a repertoire of good strategies for active pedagogy that allows you to cut back on some class preparation.
- Don't postpone course preparation until August.

Working with family, students, and colleagues:
- Talk with your spouse/partner and family about how you are trying to organize your time.
- Form a writing support group.
- Set aside regular times to spend with students and colleagues rather than having these happen by chance—perhaps one or two lunches per week; one office hour in computer lab helping students.
- Say "Let me think about it" or "May I check my schedule?" before committing to a request.
- Arrange schedule so that all work is finished before weekend; don't let work creep into family and relaxation time.
- Take one day off per week.
- Take time off when sick, otherwise stress increases and illness usually worsens.

Exercising, health, and diet:
- Exercise regularly two to three times per week, at lunch or after work.
- Take up one new hobby or extramural activity this year to relax and meet people outside your department.
- Monitor diet to make sure you are eating well.
- Schedule short exercise periods throughout the day.
- Set and keep to a regular eating and sleeping schedule.

Getting organized and keeping to schedules:
- Schedule or log time use on a daily organizer or PDA.
- Carry a small notebook, pack of post-its, or voice recorder to make quick notes before ideas are forgotten.
- Make sure daily to-do lists always include at least two small items relating to long-term goals that can be completed during the day.
- Start the day by reviewing recent accomplishments.
- Make sure daily and weekly schedules include some variety so that they don't become stale or oppressive.
- Keep and analyze a time log for 3–5 days each semester.
- Read newspapers and books as a reward for work accomplished.

Organizing the workplace:
- Reorganize your work area so that a pile of long-term projects isn't always sitting in front of you, but rather keep near you some projects you can finish in the week.
- Consolidate all of your work materials in one place rather than in several offices.
- Ensure that your work space offers some privacy.
- Organize your office and files so you can find things when you need them.

TABLE 1.1 Continued

Dealing with phone, mail, e-mail, and routine work:
- Answer e-mail during down time in afternoon rather than during productive time in morning.
- Save e-mail as a reward for finishing other work.
- Limit e-mail time; schedule it between other work; and answer only during low-energy periods.
- Don't answer the office phone. Respond to all messages once per day.
- Reduce news and web surfing and do it at times when you have less energy.
- Handle routine work in batches once a week or month.

earlier—improvements in time management do not always necessitate massive change in our schedules and lives. At the same time, time management should not be viewed as a panacea for all stress and anxiety. Thinking critically about time, priorities, and schedules is an important first step, but other types of change may be needed. For example, if a person has writer's block, regular writing periods may increase stress unless a way is found to unblock. Boice addresses these interrelated issues in his four-step program for unblocking (Boice 1997, 27–32):

1. Establish momentum through the use of free writing.
2. Establish a regimen of regular writing.
3. Establish comfort and pleasure in writing, working to avoid negative thoughts (such as confronting an unfavorable review, dealing with an unexpected and unwanted deadline, or thinking about a class that did not go well).
4. Establish social skills as a writer by seeing writing as a social process, asking for help, and gaining feedback.

The second and fourth points are familiar, but the first and third involve developing a comfort level and flow of work that highlights the quality of effort rather than just the time applied to the task.

The care we bring to our work can be as important as any change we make in our schedules. Also notable in Table 1.1 are the number of strategies that relate to health, diet, exercise, and overall physical and mental well-being. These are frequently sacrificed during graduate school and while working toward promotion. Perhaps the problem is that health and well-being often get placed in the *Important/Not urgent* quadrant of Covey's matrix and then get pushed aside by urgent tasks. The professional development literature is curiously silent on issues of health and well-being for early career academics. Yet, in my view, this is the big-picture issue that underpins this chapter and others in this book—that our health and well-being and that of those around us are important to long-term success. In some cases, this may involve adopting some of the strategies suggested in this chapter to create an effective and enjoyable balance among professional and personal responsibilities. In other cases, coaching and counseling by professionals may be the best way to reduce stress, manage anxiety, and maintain our health and well-being.

For many, however, the steps to better time management—and less stress—revolve around a few key strategies. The first is to replan schedules around Boice's concepts of moderation and balance so as to make better use of the time available during the day, week, and semester. The second is to find or create places where work can be carried out effectively and efficiently. The third is to develop a social support system to help reach time management and career goals. Too often, individuals feel they have to "go it alone" when, in fact, family, friends, and colleagues can be very supportive of plans for change. The final point, from Covey, is to take charge of our schedules by organizing time around our priorities rather than letting our agendas be shaped by events and deadlines beyond our control.

References

Ailamaki, A., and J. Gehrke. 2003. Time management for new faculty. *SIGMOD Record* 32 (2) (June):102–6. http://www.pdl.cmu.edu/PDL-FTP/stray/timemgmt.pdf (last accessed 31 August 2007).

Boice, R. 1990. *Professors as writers*. Stillwater, OK: New Forums Press.

———. 1991. Quick starters: Faculty who succeed. In *Effective practices for improving teaching*, eds. M. Theall and J. Franklin, 111–21. San Francisco: Jossey-Bass.

———. 1992. *The new faculty member*. San Francisco: Jossey-Bass.

———. 1997. Strategies for enhancing scholarly productivity. In *Writing and publishing for academic authors*, eds. J. M. Moxley and T. Taylor, 19–34. Lanham, MD: Rowman & Littlefield.

———. 2000. *Advice for new faculty members*. Needham Heights, MA: Allyn and Bacon.

Covey, S. R. 1989. *The 7 habits of highly effective people*. New York: Simon and Schuster.

Davidson, J. 2001. *The complete idiot's guide to managing your time*, 3rd ed. Indianapolis, IN: Alpha Books.

Edworthy, A. 2000. *Managing stress*. Buckingham, U.K.: Open University Press.

Ferrari, J. R., J. L. Johnson, and W. G. McCown. 1995. *Procrastination and task avoidance: Theory, research, and treatment*. New York: Plenum.

Fink, L. D. 1984. *The first year of college teaching*. San Francisco: Jossey-Bass.

———. 1988. The trauma of the first year. *Journal of Geography in Higher Education* 12 (2):214–16.

Fisher, S. 1994. *Stress in academic life: The mental assembly line*. Bristol, PA: Open University Press.

Gmelch, W. H. 1993. *Coping with faculty stress*. Newbury Park, CA: Sage.

Gray, T. 2005. *Publish and flourish: Become a prolific scholar*. Tomorrow's Professor Blog, Posting 661. http://amps-tools.mit.edu/tomprofblog/archives/2005/09/661_publish_and.html (last accessed 31 August 2007).

Hindle, T. 1999. *Manage your time*. New York: DK Publishing.

Morgenstern, J. 2000. *Time management from the inside out: The foolproof system for taking control of your schedule and your life*. New York: Henry Holt.

Pfeffer, S. 2002. Effective time management. 2002 Women in Cell Biology (WICB) /Career Strategy Columns (Archive), American Society for Cell Biology. http://www.ascb.org/index.cfm?id=1609&navid=112&tcode=nws3 (last accessed 31 August 2007).

Reis, R. M. 1997. *Tomorrow's professor: Preparing for academic careers in science and engineering*. New York: IEEE Press/Wiley.

Schouwenburg, H. C., C. H. Lay, T. A. Pychyl, and J. R. Ferrari, eds. 2004. *Counseling the procrastinator in academic settings*. Washington, DC: American Psychological Association.

Seldin, P., ed. 1987. *Coping with faculty stress*. San Francisco: Jossey-Bass.

Solem, M. N., and K. E. Foote. 2004. Concerns, attitudes, and abilities of early career geography faculty. *Annals of the Association of American Geographers* 94 (4):889–912.

Sorcinelli, M. 1992. New and junior faculty stress. In *Developing new and junior faculty*, eds. M. Sorcinelli and A. Austin, 27–37. San Francisco: Jossey-Bass.

CHAPTER 2

Career Planning
Personal Goals and
Professional Contexts

Janice Monk and Christine L. Jocoy

Maps are useful but always incomplete: exits close, freeways change, we hit dead ends. Sketching out a travel plan is important, but experienced travelers are always prepared for the unexpected. Academic life requires the same. . .. Staying open to catch the wind is a good way to turn surprises into productive outcomes. At the same time, a successful academic requires an anchor.

GALLOS 1996

Jane Gallos's travel metaphors, invoking maps and sailing, highlight important ways of thinking about academic careers: we need to clarify our values and expectations and to be well prepared but also to be open to new opportunities and changes in the contexts in which we build our lives and work. In this chapter we will consider these perspectives, offering some approaches to career planning; we will connect with other chapters such as those on balancing personal and professional lives, working toward tenure, and exploring options in interdisciplinary programs. In so doing, we consider planning along different time frames and look at the various domains of academic life and the connections among them.

In 2002, Lydia Pulsipher, professor of geography at the University of Tennessee, thinking about her own future as she came closer to retirement, convened a focus group of eleven women and men geographers of diverse ages and experiences. She sought to explore their career histories and expectations. Some commonalities emerged. Their paths had not been unidimensional or unilinear. As one commented, "a career is rarely a straight line to success. More often it is a roller coaster ride of highs and lows, of frenetic activities, of dog legs, and down times when one gets bored." The intensity of productivity fluctuates, related partly, but not only, to changes in personal life; emphases shift, though in satisfying ways. Some of the group had moved between academic and other sectors; some spent time in interdisciplinary programs or in academic administration (personal communication with Lydia Pulsipher, professor of geography, the University of Tennessee, 22 September 2006).

Pulsipher's observations point to the implications of individual development and their relations to contextual circumstances. Researchers who study personal and family changes refer to the linking of historical and social conditions to individual development as the life-course perspective (Elder, Johnson, and Conner 2003; Mabry, Giarrusso, and Bengtson 2004). It clarifies that we experience personal changes in the context of the times in which we live, the timing of life events, the relationships that link us to others, and the choices we make: that prior experiences matter for later life, but that we also grow and develop, influenced, but not confined, by what has come before. The conjunctions and synchronicities of our diverse roles have implications for our choices and the constraints within which we operate. Choices that affect professional goals arise. For example, do we establish or postpone establishing families while in graduate school and cope with what that might mean in terms of professional moves and time? (Approaches to balancing personal and professional lives are the subject of Chapter 4 by Beth Schlemper and Antoinette WinklerPrins.) Do we come as international students, torn by whether to return to careers and aging parents at home or to remain in the U.S.?

Life-course perspectives identify important cohort effects that we may not recognize ourselves, nor may those who advise and evaluate us. The large group of academics retiring in the early twenty-first century, for example, often those who have advised recent and current doctoral students, entered the labor market when universities were expanding dramatically (partly to meet the needs of the baby boomers entering college in the late 1960s and partly as a result of expanded federal funding for science linked to Cold War politics). Over the intervening decades, the size and nature of academic labor markets have waxed and waned in response to economic, political, and demographic trends, with implications for those seeking to enter the professoriate. In recent years, pressures on university budgets, for example, have seen an increase in part-time and non-tenure-track positions in universities (Purcell 2007), while growing support for interdisciplinary endeavors fosters the creation of shared appointments across units. Developments in technology shape demands for particular skills within higher education and create new opportunities outside academia. Being aware of changes in contexts and of life-course perspectives helps us to think about priorities and options (Monk 2001).

CLARIFYING VALUES AND EXPECTATIONS IN THE CONTEXT OF PROFESSIONAL RESPONSIBILITIES

Building on life-course perspectives, aspiring academics should ask themselves an array of questions, such as the following, to help form career plans. It is also necessary to revisit those questions as one moves along:

- How are my personal and professional lives interdependent?
- Is my career linked with that of a partner/family?
- How do policies and practices of institutions or organizations constrain choices?
- How might I/we negotiate within the institution or organization? How might I work to change it?
- What do I do well? How can I identify my strengths? What can I do to improve my qualifications?
- How flexible/broad/specialized do I want/need to be?
- With whom do I wish to communicate?
- What are my expectations and desires with regard to contributing to institutions, organizations, and society?

In reflecting on such questions, there are several possible paths to explore. First, reading autobiographical studies by other academics helps to clarify goals and draw road maps, to see how others have found ways to navigate through the waters, making adaptations and working to change institutions (Frost and Taylor 1996; Archer 2001; Purcell 2007). Kevin Archer, for example, recounts the ups and downs of attempting to combine his graduate education and early career years with his wife's education and job searches and an outcome that, for him, ultimately brought satisfying changes in his aspirations, research agendas, and approaches to teaching. A second valuable source of reflections that can assist early career professionals to think through their goals and strategies are the columns in *The Chronicle of Higher Education*. These offer first-person accounts of many challenges that academics face and ways in which they address them. Third, after general reflection, it is helpful to try to articulate answers in writing. One of us, Christy, has adopted this approach, preparing a self-assessment and then proceeding to create a development plan (see below). The other,

Jan, has given the task to graduate students in a professional development seminar, asking them to share their responses with one other student and then to discuss the experience of trying to engage in such an exercise. It seems that sometimes the students de-emphasize their strengths in completing the assessment, and we encourage aspiring academics to think of strengths and talents as well as limitations!

Before providing a concrete example of career planning in practice, we discuss the importance of clarifying values and expectations in two areas: finding employment and balancing service with other professional responsibilities.

IDENTIFYING PROFESSIONAL OPPORTUNITIES

Opportunities take multiple forms. Clearly, a high priority for most will be to identify a faculty position. But there may be other options for you to consider, such as seeking a postdoctoral fellowship to extend your research preparation. Beyond an appointment, other professional activities also are a part of the career path, for example, seeking service roles in professional organizations that will extend networks, publishing one's research, or seeking out possibilities for interdisciplinary collaboration. We will address first the theme of securing a faculty position, then move on to the place of service in professional life. Other chapters in this book take up aspects of publishing (see Chapter 13 by Stanley Brunn) and interdisciplinary work (see Chapter 14 by Craig ZumBrunnen and So-Min Cheong).

An array of advice manuals and columns deal with locating and securing academic jobs, examples of which are noted in this chapter. Discipline-based professional societies usually offer relevant materials on their web sites and in print. They cooperate with employers to advertise positions and host job interviews at professional meetings. Some have data banks where aspiring professionals can post their credentials. *The Chronicle of Higher Education* features advertisements from many institutions and in diverse fields. It also publishes columns that offer informal career advice. It is especially useful for locating the increasing array of positions that are in interdisciplinary programs and that may not appear in disciplinary sources. Advertisements for many academic positions are distributed on listservs with the aim of reaching candidates in particular subfields. Individual academic departments often offer career information on their web sites and circulate announcements internally.

Once possible positions have been identified the application process becomes intensive. Resources are available to support this task. One of the most useful is the book by Heiberger and Vick (2001), which addresses critical issues such as planning and timing the search, preparing written materials (vitas, letters of application, ancillary materials, web sites), interviewing, negotiating, and others. Columns in *The Chronicle of Higher Education* in which individuals reflect on their own experiences (often with humor) are also helpful. It pays to become familiar with such materials well before the last minute.

For many aspiring academics, the initial hope is likely to be securing a tenure-track position, though personal circumstances and labor market conditions at times make that goal elusive (see Chapter 5 by Susan Roberts). Faculty members in doctoral departments are not always well equipped to offer advice about alternatives to academic careers. This is one reason why professional associations are seeking to enhance the advising materials they offer. Important supplements include networking with professionals outside the department where one is studying and attending presentations and job fairs at conferences.

Learning about alternatives is important not only for the individual seeking a position but also for future professors to guide students about their options. Despite being well prepared and skilled in search processes, some doctoral graduates seeking academic careers nevertheless find themselves in year-at-time positions with substantial teaching responsibilities and constant new preparations, not what they had anticipated. It is important then to consider that professional careers in government agencies, private companies, and the nonprofit sector can offer rewarding alternatives. Furthermore, some who take such positions find opportunities to maintain academic ties and offer valuable professional mentoring to graduate students by combining part-time adjunct teaching with their other employment. Resources are available to work with such options and also to learn of their problems (Greive 1983; Dubson 2001). Dual-career couples also face particular challenges that institutions are increasingly recognizing and seeking ways to accommodate (Wolf-Wendel, Twombly, and Rice 2003).

Recognizing and seeking alternatives need not just be a "last ditch" strategy. Making connections with faculty and students in interdisciplinary programs and research centers while a doctoral student (for example, in courses or assistantships) can widen your knowledge of options for employment, as well as offering preparation for positions in interdisciplinary programs. Taking such initiatives can also enhance your qualifications for meeting the increasingly common expectation of funding agencies that scholars work in teams and across disciplinary lines.

THE PLACE OF SERVICE IN PROFESSIONAL LIFE

In the previous section we emphasized finding a professional position. We now turn to an aspect of experience over the career that is not always considered part of the formal curriculum of graduate preparation with its focus on learning to be a researcher under the guidance of an advisor and a committee. Graduate education also offers experience and preparation for the teaching roles of the academic, as by way of teaching assistantships, mentoring by faculty and fellow graduate students, or workshops offered by a university teaching center or within a professional association. But the third component of academic work, service, is often not discussed in graduate school; further, early career faculty are often advised "service doesn't count." We disagree with that position, as does Donald Hall (2002) in an extended autobiographical meditation on his career in academia. It depends on what is meant by "counts." In our experience, service is a valuable way to learn about your profession and institution, to support your research and teaching, and to guide you in thinking about one of the questions we posed above: "What are my expectations and desires with regard to contributing to institutions, organizations, and society?" We have therefore chosen, in this section, to address the place of service in faculty life, especially in relation to other aspects of professional development.

The functioning of academic workplaces and society, as well as of individual careers, depends on people who contribute service. Manuscripts and proposals need to be evaluated, letters of recommendation written, conferences organized, the work of interest groups facilitated, journals and newsletters edited, web sites and listservs maintained, candidates for positions and promotions evaluated, curricula developed and revised, visitors hosted, students advised, student organizations sponsored, outreach made to K-12 schools and the public to communicate and sustain your field. The list goes on. Some see such work as a sink of time and energy, a diversion from the real work of research and teaching. But it is essential for many reasons, some altruistic, others that are vital in supporting an academic's career, psyche, and the health of departments and institutions. Service is the glue that holds research and teaching together (Hanson 2007). Without it, support of research and its dissemination would not flourish, nor would the climate for teaching and learning. Getting to know how an organization operates and meeting people beyond one's immediate circle who can offer support and feedback on work and progress are just some of the benefits of service. It can also enrich personal life. For the new faculty member, especially the young, single person, the loss of daily contact with peers that characterized graduate school can bring feelings of loneliness and isolation. Developing relationships with colleagues at the local, national, and international levels through service builds connections to those with shared research and teaching interests and sustains rich, personal friendships.

Reflections by some geographers who made service contributions as graduate students reveal diverse approaches and benefits for their professional development. Kim Eisele initiated the student-edited magazine *You Are Here: The Journal of Creative Geography* (http://www.u.arizona.edu/~urhere) with the goal of communicating with publics beyond academia. She reported learning

> tons, (most of all) how to be a better leader. . . . I have discovered my many mistakes in motivating (or failing to motivate) other students. This was really my biggest struggle, getting the idea out of my head and enough into the hands of others that they could

feel some ownership. Overall, I learned a lot about management, opportunities, organization, grant writing, public relations and how to channel creative activity. (Monk 1999, 288)

Susan Mains, who as a graduate student maintained two substantial listservs in her discipline for four years, has written,

it has played a crucial role in making me feel part of geography, and particularly part of a community of feminist/critical geographers. The role has taught me that diplomacy is important, but so too is the ability to be assertive and keep open, as much as possible, spaces in which a variety of geographic perspectives can be discussed. (Monk 1999, 288)

Their comments are reminders that research, teaching, and service are interdependent. Being the graduate representative on a committee in a professional organization can offer introductions to colleagues who will review draft manuscripts or who may be called upon when time comes for tenure review. Working with a local graduate student organization, serving on a departmental curriculum committee, or helping to host visitors will widen networks and professional understanding of academic processes and may lead to new friendships. But the key in service activities is reciprocity, to contribute as well as to receive. To return to our road metaphor, service is a two-way street. We encourage exploration of how it connects and contributes to other parts of one's professional and personal lives.

CAREER PLANNING IN PRACTICE: SELF-ASSESSMENT AND DEVELOPMENT ACTION PLANS

As the quote at the beginning of the chapter states, "Sketching out a travel plan is important. . ." (Gallos 1996). It is important for a variety of reasons, and we provide just a few here. First, different people and institutions will set itineraries and priorities for you. For example, the focus of advice for academics is so often on how to do enough to get tenure with little consideration for balancing work with other life ambitions. It is worth taking some time to articulate when personal and professional values and goals are in agreement and when they diverge. Second, the demands of daily life and responsibilities make it hard to find the time to make plans. Nonetheless, it is important to assess and reflect on one's current position and future goals with at least some regularity, so as to create reference points or benchmarks for stages of professional development, especially for those days when you wake up and feel like you have no sense of your current situation or how it might change. Setting aside the time to write periodically about values and goals can facilitate such reflection and serve as a reminder of its importance.

Self-assessment and *development action plans,* then, are some techniques for professional development. These techniques are adapted from conventional personnel evaluation practices used in most businesses, including universities and colleges. However, in contrast to professional development plans focused solely on job-related requirements and goals, the planning techniques described below and in the accompanying Activities 2.1 and 2.2 are meant to promote reflection on professional and personal development beyond the periodic reviews or evaluations required by employers.

Self-Assessment

The purpose of self-assessment is to reflect on and assess one's own values, interests, strengths, and weaknesses to gain insight into the relationship between career and other aspects of life and to plan for the future accordingly. Donald Hall (2002) encourages us to view self-assessment as a reflexive process of routinely creating and sustaining one's professional self-identity. It "entails a willingness to articulate one's values and priorities, a willingness to engage critically and openly one's sense of what an [academic professional] 'is' and 'does' " (Hall 2002, 10). Furthermore, it requires "movement between and an awareness of personal agency and an acceptance of where and how agency is limited" (p. 10). When articulating values, for example, it is important to examine one's own acceptance of professional norms and become educated about the consequences of individual decisions that adhere to or depart from them. Aspiring academics should ask themselves how they measure their own professional successes and priorities. Is the goal to become a top researcher at a research-intensive university? Is the goal to work as a consultant in a non-academic setting? Is it to teach and serve students who are the first in their families to go to college? In addition, these goals should be put in context by asking questions about how one's values, interests, strengths, and weaknesses relate to achieving them. You can engage with these ideas by completing the self-assessment exercise in Activity 2.1.

Development Action Plan

The purpose of an action plan is to translate the values, interests, strengths, and weaknesses identified in reflexive activities such as self-assessment into specific action steps. It involves defining professional performance expectations in conjunction with evaluations of what it will take to meet those expectations while remaining committed to broader personal goals and values. Such a task is a balancing act that requires thinking through a course of action. Some suggested steps for completing an action plan follow (and you can develop your own plan by completing Activity 2.2).

The first step consists of setting a planning time frame. A plan does not have to be comprehensive; it can be completed for short time frames and for different areas of professional work. The second step involves identifying specific performance expectations based on both personally driven goals and those of your employer. For instance, if you value creating research opportunities for students, it is useful to think through how to meet research and publishing expectations set by your institution in the context of creating hands-on learning experiences for students. The third step is to identify the knowledge, skills, and behaviors needed to achieve each performance expectation. In the case of new university faculty, they need to develop knowledge of retention, tenure, and promotion criteria. A fourth step is to determine whether formal professional development or training exists within or beyond your institution that can be accessed for assistance in meeting a performance expectation. For instance, your institution may hold grant-writing workshops. A fifth step involves articulating the resources and support needed from others to fulfill an expectation (e.g., funding, mentoring, social support, collaboration, or release from teaching a course). A last step is to establish target deadlines for achieving performance expectations.

It is important to keep self-assessment in mind when setting action steps, especially in terms of setting priorities and boundaries that allow for balancing personal values with professional ambitions. One approach is to think about setting expectations that are consistent with values and interests. In addition, the creation of a plan should not suggest that reflection is a discrete event. Reflection should occur continuously in the daily practice of professional and personal development. As Schön's (1983) research on professionals indicates, a key component of the practice of professions is thinking about what one is doing while one is doing it and reflecting on past and current practices so that one can adjust to unanticipated future situations that diverge from career plans.

PREPARING SELF-ASSESSMENT AND ACTION PLANS: ONE ACADEMIC'S EXPERIENCE

The narrative below illustrates how one of us, Christy, prepared a self-assessment and development action plan for two areas of performance expectations: teaching and service. For the self-assessment, the time frame is a three-year period, commencing with the beginning of a tenure-track faculty appointment (Figure 2.1).

Christy's Self-Assessment

Like many new academics, I was concerned about balancing work responsibilities, career aspirations, and a social life in my new home, especially since I moved across the country to take my position. Realizing that establishing this kind of balanced integration and matching of my values to circumstances would take several years, I selected a multiyear benchmark.

Through the self-assessment, I attempted to articulate my priorities to create a mental picture of what I would like to have accomplished or be

FIGURE 2.1 Self-Assessment for Christine L. Jocoy, Assistant Professor, California State University, Long Beach, Fall 2004[*].

Complete this section to describe your development priorities for the **next 3 years**. Considering the following information:
- Your personal career goals
- How these goals connect with your personal life
- The goals of your place of employment

Planning period: From **Sept 04** To **Sept 07**

1. List the values you bring to relating your career to other aspects of your life.
   - While I am willing to work hard and long hours to achieve specific goals, I do not want to be a workaholic.
   - I want to be physically fit and pursue interests/hobbies unrelated to my work.
   - In addition to developing friendships within my university, I want to have friends outside of academia.
   - I want to know and participate in my local community.
   - I want to have a diverse set of mentors on whom I can rely for advice.

2. List work-related strengths and interests that you would like to build on.
   - Writing skills—build on previous publishing success.
   - Collaborative research—find other researchers at University
   - Community engagement—develop research projects affiliated with service to local community
   - One-on-one student advising—engage students in research

3. Identify ways you would like to contribute to your profession/organization beyond the immediate responsibilities of your position.
   - Learn Spanish—lots of my students are native speakers
   - Join University Commission on the Status of Women
   - Participate in activities of the University Multicultural Center
   - Participate in Alternative Spring Break trips for students

4. Describe areas for improvement in terms of the responsibilities of your current position.
   - Develop courses, interactive assignments, active learning activities
   - Grant writing
   - Develop new research projects
   - Engage professional networks through Association of American Geographers (AAG) specialty groups

(Continued)

FIGURE 2.1 Continued

---

5. Describe areas for development in terms of future professional responsibilities.
- Participate in AAG specialty groups
- Participate in international geography networks—find international teaching and research exchange opportunities

*Thanks to Ann Taylor of the John A. Dutton e-Education Institute at The Pennsylvania State University for providing a model on which the Self-Assessment and Development Action Plan forms are based.

---

working toward at the end of three years. This included recognizing not only what I need to do to fulfill work responsibilities (e.g., designing and teaching new courses, publishing, applying for grants, and developing a research agenda), but also what I want to do to improve my professional strengths (e.g., finding collaborators for research and writing, engaging students in research, and participating in local community-based research) and live up to my personal values (e.g., learning about and contributing to the local community, developing social networks of friends and colleagues, and balancing work and personal life). While reflecting on my priorities, I tried to keep in mind Donald Hall's advice: "There is no one career path that guarantees success nor one set of accomplishments that signifies success" (2002, 11). Admittedly, my self-assessment does not reflect an overwhelming ambition to be a leading researcher in my field, in spite of the messages received from my Ph.D.-granting institution. As the primary mission of my university is teaching, I am not under institutional constraints to direct most of my attention to research. However, my professional goals include continuing engagement with active and contemporary academic research and keeping open the possibility of moving to a new university if my goals change.

Reflecting on these priorities two years after first articulating them, my first reaction is a sense of accomplishment that I achieved some of these goals or at least initiated activities related to meeting them. It is important to reflect on and be able to articulate accomplishments, not only for employment reviews, but also for personal fulfillment (especially given that there is always more to do!). The document also reminds me of the things I have not yet pursued and the things that I want to pursue more intentionally. Some goals can be pursued within a longer time horizon (for example, learning Spanish and participating on the University's Commission for Women). Others I want to spend more time on (for example, maintaining social networks and exploring my community). I can feel confident doing this because I have met many of my work responsibilities (for example, getting a few publications accepted and receiving a grant for new research). Such realizations help to ground me, especially when the demands of daily activities produce feelings of being directionless and distracted from my values and broader goals. Rereading the self-assessment helps me to feel more confident about the choices I have made and the reasons why I made them.

Christy's Development Action Plan

In my development action plan (Figures 2.2 and 2.3), I illustrate plans for meeting two different performance expectations: one related to teaching and the other to service. In the first example (Figure 2.2), I created an action plan to improve some teaching materials. This relates to broader performance expectations for my teaching responsibilities, such as designing courses and receiving decent scores on student evaluations, but the narrow focus on a few specific improvements makes it easier to demonstrate progress and achieve it at a manageable, if incremental, pace. Incremental plans may help relieve the pressure one faces as a new faculty member in managing (and limiting) the time spent on teaching preparation.

Furthermore, the implementation of this plan relates to several of the goals identified in my self-assessment. First, to develop exams and other assignments, I established relationships via professional associations for sharing course materials with human and economic geographers at other universities. In addition to fulfilling a teaching expectation, this action step met my goal of engaging professional networks. Second, my interest in creating materials to teach proper citation style and plagiarism avoidance, which began when I was a graduate teaching assistant, led to collaborative research with a former graduate school advisor and ultimately a publication in an education journal (Jocoy and DiBiase 2006).

In the second example (Figure 2.3), I address performance expectations for service, specifically by attempting to select service activities that help further my integration into the college and university and to combine research and service expectations. When I began my position in 2004, the departmental guidelines for Retention, Tenure, and Promotion (RTP) had not been updated in an official document since 1997, even though the college guidelines for RTP had been revised and made more specific since then. Knowing that I would soon be evaluated based on these documents, I volunteered for my department's personnel logistics committee, which was responsible for revising the RTP guidelines. This service assignment allowed me to become intimately familiar with the department's performance expectations while at the same time fulfilling my service expectations. I found it worthwhile to be proactive about selecting service commitments, thereby maximizing the benefits from the time spent on service.

A second way in which I combined performance expectations with self-assessment goals was to meet RTP service requirements while engaging my interest in community service and individual student mentoring. I proposed and received grant support for research that included student researchers and cooperation with community planners and local government officials. Participating in the City of Long Beach's community planning initiative to

FIGURE 2.2 Development Action Plan

Planning Period: From Dec 2003   To   Dec 2004
Review and update this plan periodically.

Performance Expectations for Teaching	Knowledge, Skills, and Behaviors Needed to Achieve Each Expectation	Professional Development Activities	Resources and Support Needed from Others (e.g., dept, univ., professional assoc., grant providers)	Target Dates for Expectations
1. Two classes (World Regional and Economic).	1. Find and read appropriate texts, design exercises, write lectures.	1. Attend Univ summer teaching workshops.	1. Obtain release from one course through Dean's office	1. Dec 2004
2. Earn decent student evaluations.	2. Identify students' needs.			2. Dec 2004
3. Create a bank of exam/quiz questions appropriate for intro human geog courses.	3. Collect exams from professors who have taught human geog.			3. Aug 2004
4. Create materials for teaching proper citation style and ways to avoid plagiarism.	4. Revise existing citation and plagiarism materials from courses for which I served as a teaching assistant.		4. Access Univ subscription to online plagiarism detection service	4. Sept 2004
5. Establish an online, automated system for generating quizzes to evaluate student progress.	5. Use Univ online course management system to design tool.	5. Work with Univ technical support staff.		5. Dec 2004

FIGURE 2.3 Development Action Plan

Planning Period: From Aug 2004   To   May 2005
Review and update this plan periodically.

Performance Expectations for Service	Knowledge, Skills, and Behaviors Needed to Achieve Each Expectation	Professional Development Activities	Resources and Support Needed from Others (e.g., dept, univ., professional assoc., grant providers)	Target Dates for Expectations
1. Department committee service—Personnel logistics	**1.** RTP guidelines for Univ, college, dept	**1.** Univ and College sponsored RTP workshops		**1.** Sept 2005
2. College/Univ committee service—College Faculty Council	**2.** College of Liberal Arts policies			**2.** Once a month meetings through May 2005
3. Service to profession		**3.** AAG specialty group newsletters and business meetings at annual meetings; AAG listservs		**3.** Dec 2005
4. Univ and Community service—City of Long Beach 10-year plan to end homelessness	**4.** Recruiting student participants from Geography Student Association; Housing and homelessness policy—federal, state, local		**4.** Grant money for student research assistants; agreement of participation of city officials in research and service activities	**4.** May 2005

address homelessness enabled me to conduct participant observation research in urban social and economic geography while at the same time performing community service and advising students involved in research. Creating and reflecting on development action plans enabled me to consolidate efforts at meeting performance expectations for service and research with personal goals of participating in the local community and meeting people outside of academia.

In hindsight, I probably do not reflect on these plans as frequently as I should. Nonetheless, they are useful for preparing annual performance reviews and justifying decisions to decline requests for additional service participation or involvement in new research projects, for example. They are especially helpful for thinking through whether to pursue new opportunities that arise. Does this new opportunity fit into my existing plan? Is it important enough for me to alter my plan? Which objectives have I completed such that I have time to go in a new direction? By reflecting on the plans, I can answer more confidently important questions about how to maintain a balance between performance expectations and other professional and personal goals. Furthermore, I am comforted when reminded of how much I have accomplished, especially when it feels like the demands on my time seem to prevent me from finishing any project in a timely manner.

BRINGING IT ALL TOGETHER

In this chapter, we have aimed to illustrate how personal and professional goals and strengths can be brought together and pursued. We highlight the importance of reflection and planning, linking self-assessment with the articulation of specific developmental goals and activities. In so doing, we pay special attention to the synergy among service, teaching, and research and to a perspective on the life course that sees career development as a gradual process, one that will shift over time as our aspirations, contexts, opportunities, and skills evolve. We encourage you to complete your own self-assessment and development plans, to reflect on where you hope to travel as well as how far you have come.

References

Archer, K. 2001. Through a glass darkly: Re-collecting my academic life. In *Placing autobiography in geography*, ed. P. J. Moss, 62–77. Syracuse, NY: Syracuse University Press.

Dubson, M. 2001. *Ghosts within the classroom: Stories of college adjunct faculty—and the price we all pay.* Boston: Camel's Back Books.

Elder, G. H., M. K. Johnson, and R. Conner. 2003. The emergence and development of Life Course Theory. In *Handbook of the life course*, eds. J. T. Mortimer and M. J. Shanahan, 3–19. New York: Kluwer Academic/Plenum.

Frost, P. J., and M. S. Taylor, eds. 1996. *Rhythms of life: Personal accounts of careers in academia.* Thousand Oaks, CA: Sage.

Gallos, J. 1996. On becoming a scholar: One woman's journey. In *Rhythms of life: Personal accounts of careers in academia,* eds. P. J. Frost and M. S. Taylor, 11–19. Thousand Oaks, CA: Sage.

Greive, D., ed. 1983. *Teaching in college: A resource for adjunct and part-time faculty.* Cleveland, OH: Info-tec.

Hall, D. E. 2002. *The academic self: An owner's manual.* Columbus: The Ohio State University Press.

Hanson, S. 2007. Service as a subversive activity: On the centrality of service to an academic career. *Gender, Place and Culture* 14 (1):29–34.

Heiberger, M. M., and J. M.Vick. 2001. *The academic job search handbook,* 3rd ed. Philadelphia: University of Pennsylvania Press.

Jocoy, C., and D. DiBiase. 2006. Plagiarism by adult learners online: A case study in detection and remediation. *International Review of Research on Open and Distance Learning* 7:1. http://www.irrodl.org/index.php/irrodl/article/view/242/466 (last accessed 30 August 2007).

Mabry, J. B., R. Giarrusso, and V. L. Bengtson. 2004. Generations, the life course, and family change. In *The Blackwell companion to the sociology of families,* eds. J. L. Scott, J. Treas, and M. Richards, 87–108. Malden, MA: Blackwell Publishing.

Monk, J. 1999. Valuing service. *Journal of Geography in Higher Education* 23 (3):285–89.

———. 2001. Many roads: The personal and professional lives of women geographers. In *Placing autobiography in geography,* ed. P. J. Moss, 167–87. Syracuse, NY: Syracuse University Press.

Purcell, M. 2007. "Skilled, cheap, and desperate": Non-tenure-track faculty and the delusion of meritocracy. *Antipode* 39 (1):121–43.

Schön, D. A. 1983. *The reflective practitioner: How professionals think in action.* New York: Basic Books.

Wolf-Wendel, L., S. B. Twombly, and S. Rice. 2003. *The two-body problem: Dual-career couple hiring policies in higher education.* Baltimore, MD: Johns Hopkins University Press.

CHAPTER 3

Developing Collegial Relationships in a Department and a Discipline

M. Duane Nellis and Susan M. Roberts

One of the most personally enjoyable aspects of an academic career may be developing close relationships with colleagues. Collegiality, a term with roots deep in religious and educational history, refers to the nature of these relationships and has come to have a generally positive meaning. Collegiality is valued by most academics, and by most institutions in which graduate students and faculty members work (Bennett 1998; Fogg 2006). Developing positive relationships with colleagues with whom you interact on a day-to-day basis is an important aspect of your professional life. While being a "good citizen" of the academic units you are affiliated with—be it a program, a department, a college, or a center—is a significant part of your professional development, it is not often explicitly discussed. Rather, collegiality is often treated in a taken-for-granted way and left as a hazy yet somehow important element of an academic career.

In this chapter (in line with recent literature in professional development), we discuss how collegiality is understood in academe, and we give some practical ideas about how to actively develop collegial relationships—not just with those people you see in the hallways everyday, but also with more distant colleagues. After setting the scene, in terms of some of the major changes underway in higher education, we move to discuss strategies for developing collegial relationships with others in your department, and then beyond the department. We discuss collegiality in terms of a department and a discipline, but our points should be broad enough to apply to other institutional settings too.

A recent headline in *The Chronicle of Higher Education* proclaimed "Young Ph.D.s Say Collegiality Matters More Than Salary" (Fogg 2006), and indeed most academics do prize collegiality, but it doesn't just happen. Collegiality is present when individual faculty and administrators commit to nurturing it. However, while collegiality is seen as a definite positive, whether it should be a factor in tenure evaluations is a matter of fierce debate. Even if it is not explicitly given as a criterion, collegiality undoubtedly comes into play (Phelps 2004). As Catharine Stimpson, in responding to a high-profile report by a task force of the Modern Language Association in the US (MLA 2006) that took up this issue, noted: "Of course, 'collegiality' should not be an explicit criterion for tenure, because it might reward the good child and punish the up-start. However, . . . because tenure is forever, at least on the part of the institution, it is legitimate to ask how a candidate will contribute to the institution's long-term well-being" (Stimpson 2006). The debates over the value and fairness of using collegiality as a factor in evaluation are ongoing, and we return to them in our conclusion.

COLLEGIALITY IN TODAY'S HIGHER EDUCATION ENVIRONMENT

This is a time of great changes in higher education. Students of today have different expectations regarding learning environments and styles. Technological innovation is dramatically changing the ways in which we interact with students, as well as with our departmental colleagues and others across the university and beyond. Many of the major research questions are at the interface of disciplinary boundaries or link to questions that span across broad disciplines such as geography. Faculty members are also expected to be more focused on new modes of assessing student learning and outcomes-based learning. In addition, serving students from beyond the immediate campus boundary is creating new classroom dynamics including virtual classroom settings. At some institutions, faculty scholarship for tenure and promotion is changing in ways that allow for more flexible profiles that maximize faculty talent and performance in areas of scholarship that extend to teaching, civic engagement, and outreach (Kennedy 1997; Duderstadt 2000). In the context of these substantial changes, working in a cooperative, constructive environment of engagement while embracing what is best in these dynamic changes will certainly contribute to one's success.

Within this new higher education environment, citizenship is becoming increasingly important as a criterion for tenure (see Chapter 5 on "Succeeding at Tenure and Beyond" by Susan Roberts). Even with strong credentials, a faculty member may be refused tenure if he or she is perceived as being malevolent, uncooperative, or uninterested in working with colleagues within a department (Weeks 1996; Lewin 2002; Diamond 2004; Phelps 2004). We return to this controversial matter in our conclusions. Suffice it to say, though, that anyone seeking tenure needs to be aware of the importance of collegiality and be cautious about appearing unwilling to compromise in adapting to changing expectations in today's dynamic education environment.

CONTRIBUTING TO COLLEGIALITY AT THE DEPARTMENT LEVEL

Few areas of academic life are more central to one's success than collegiality and working in a community of scholars. Collegiality plays a pivotal role in academia, lying at the intersection of different aspects of academic life (Åkerlind and Quinlan 2001; Silverman 2004; Fogg 2006). Research has found that support of faculty colleagues was closely associated with many of the intrinsic rewards that are the mainstay of an academic career (Olsen and Sorcinelli 1992). In many cases, departments will assign an appropriate faculty mentor to work with junior faculty in an effort to enhance collegiality. Encouraging early career faculty to be more engaged and networked is increasingly regarded as central to a department's faculty development and mentoring programs. Often, department chairs are seen as important advocates and supporters of junior faculty as they work toward establishing a constructive work environment based on a strong sense of collegiality (Sorcinelli 1992). Common elaborations of this theme include mentoring, networking, providing a sense of support or belonging, and encouraging junior professors with their research proposals and writing projects (Jarvis 1992). Just as departments, through department chairs and faculty, can create environments that help lead junior faculty toward a more collegial environment, the junior faculty members themselves can demonstrate certain characteristics of interaction that are often perceived as signature traits of a collegial colleague. Although there are many books that focus on relevant issues such as effective communication and leadership traits, the following represent some central themes that we feel are most helpful to being perceived by colleagues as collegial.

Knowing oneself is often considered central to knowing others and building community. As Margaret Wheatley (1992) has underscored, "power in organizations is the capacity generated by relationships." Robert Boice, drawing upon his many years of detailed research on early career faculty, points out that new faculty "... agree on one thing: no matter how much they value their autonomy as professors, they still rely most on colleagues for success as teachers, as productive researchers, and as contented professionals" (Boice 1992, 19). Boice's research reinforces the point that, as faculty members, not only must we each commit to effectively teach our courses, conduct our scholarship well, and perform adequate service, we must also work just as intensively toward building constructive relationships with colleagues—in part because doing so will increase our own likelihood of succeeding in and enjoying our professional lives (see also Boice 1991, 2000). Building constructive relationships can be enhanced by

1. fostering and maintaining open communication with colleagues,
2. treating colleagues with respect,
3. having a vision for where your career is headed and being sensitive to how it aligns with others in the department and university,
4. maintaining integrity in dealing with colleagues,
5. being optimistic—focus on the glass being half-full not half-empty,
6. being a good listener, and
7. being strategic in one's thinking (Nellis 2006).

We would like to elaborate briefly on each of these points. It is important to have open dialogue with other faculty. Seek their advice on research ideas or teaching strategies, gain their insights about campus resources or ask them for ideas about solving problems. Taking a proactive approach to open communication will only be perceived as positive in the context of collegiality.

Treating colleagues with respect is also essential. Just because other faculty have scholarly interests that may conflict with your point of view doesn't mean you should openly challenge your colleagues or try to embarrass them. On a related note, it is best to avoid talking about departmental colleagues to undergraduate students, graduate students, or other faculty in ways that might be perceived as demeaning.

Having a vision as to where you want your career to head and being willing to look toward aligning your vision so it complements the department's and university's vision is also important (Chapter 2 on "Career Planning" prompts you to think about your values and offers approaches to developing a professional and personal plan, while Chapter 4 on "Balancing Personal and Professional Lives" includes some issues that relate to collegiality). If you, for example, insist on being allowed to teach a particular class regardless of departmental interests and needs, this can be perceived as less than collegial, and contrary to what is best for the overall good of the department.

Honesty and integrity are crucial elements to developing a collegial atmosphere with departmental faculty colleagues. Obviously, this doesn't necessarily mean you should be "brutally honest" when kindness would suggest you keep some thoughts and comments to yourself. Nonetheless, colleagues will expect you to be open, not misleading, and they will respect and appreciate you for delivering when you say you will on departmental tasks.

Although sometimes difficult, collegiality is often enhanced when you are perceived as positive, not negative. This doesn't mean glossing over real problems or issues, but many challenges one faces in an academic career can be dealt with best by adopting a positive approach based on engaging the problem rather than a negative one based on simply complaining about it. As academics, we are trained to be critical and analytical, but as we apply these characteristics we need to be careful and constructive in our approach to problem-solving.

Being a good listener is also sound advice as one works toward being perceived as more collegial. Faculty colleagues and department heads may provide mentoring advice or helpful ideas related to research or teaching. Quality listening and follow-through will certainly help create a more positive interpersonal work environment.

Focusing on your own development as a faculty member—but in the context of the plans of your department and the goals of your colleagues—will also be perceived favorably as a contribution toward collegiality.

Taking an active role in working toward being more engaged with faculty colleagues through a commitment toward collegiality often has many rewards toward faculty success. At the same time, such efforts do not abrogate one's responsibility to perform duties in scholarship, teaching, and service, consistent with expectations to be successful in one's academic career. Yet, taking a proactive approach toward collegiality will, in many ways, help the aspiring faculty member meet these very expectations and reach his or her own goals, while at the same time such an approach can help create a more positive and less stressful work environment for all members of the department (Boice 1991).

Having said all this, unfortunately many faculty find themselves in academic units characterized not so much by collegiality as by dysfunctionality. What if you find yourself in a department with colleagues who are obstructive, rude, poor listeners, and/or disengaged? How can you build collegial relationships in a place where the culture of collegiality has not been fostered and valued? Certainly, almost every academic has either first or second hand "war stories" of the tremendous difficulties faced by junior faculty in dysfunctional or otherwise unsupportive settings. In such circumstances, developing collegial relationships with difficult departmental colleagues calls for particular strategies building upon the points above.

For instance, regarding the fifth point, Robert Boice's research has confirmed that faculty who succeed are those who do not dwell on criticism of their work or on negative experiences (Boice 1991, 1992, 2000). Instead, Boice points out that successful faculty were distinguished by being able to maintain an optimistic outlook even in circumstances that drove others toward pessimism and cynicism. In part, as his research shows, being optimistic is related to having social and support networks that cumulatively contribute to the faculty member's productivity (Boice 1991). So, while a faculty member's immediate academic setting may present some serious challenges to a junior colleague's attempts to build collegial relationships, maintaining optimism is important. Thinking strategically about how to develop collegial relationships even with the most difficult of colleagues may also prove effective. For example, even though a senior colleague may be unfriendly and appear generally disengaged, is there perhaps some area of expertise or experience he or she has about which you could seek advice? Perhaps they are very familiar with the campus map library and could be asked for advice on using the library's resources, maybe for a class exercise. Or, perhaps a usually grumpy senior colleague has led many field trips and could be approached for any tips he or she may wish to share in that regard. Initiating such conversations can be the beginnings of respectful, if limited, engagement. In some cases, collegiality might entail connecting with colleagues regarding some particular interest of theirs. If they have a poster of a sports team in their office, asking how he or she became a fan of that team (even if you don't know anything about or care about the sport or the team!) can be a friendly gesture. If you are new in town, asking about good parks to visit, nice restaurants, or reliable places to get a bike or a car fixed can again be preludes to cordial relations with colleagues.

In addition, sometimes senior faculty may attempt to enroll new faculty members in political struggles within the department. Knowing about any long-standing battles can be important, but getting involved in what are essentially other people's struggles is unwise. Establishing yourself as an optimistic, cheerful person who makes attempts to engage each colleague, regardless of which "side" of this or that battle they may be on, is surely a challenge—but one worth taking on.

Most academics don't find it easy to, as one might put it, "schmooze," and, for many, joining in whatever social events happen in the life of a department may not be comfortable. Even so, try to express interest in colleagues and their lives by engaging in hallway chat and the small gestures to start conversation discussed above (see Phelps 2004). Meanwhile, even if the immediate academic home presents such challenges to an early career faculty member, academe presents other, usually less stressful opportunities to build supportive networks and sustaining professional relationships. It is to these that we now turn.

BUILDING COLLEGIAL RELATIONS BEYOND THE DEPARTMENT

Cultivating a network of colleagues beyond your department with whom you share interests is rewarding even if you have strong and positive relationships with departmental colleagues (Hanson 2000). Peers and mentors from a graduate program may be one set of colleagues with whom beginning faculty members will enjoy keeping in touch, but expanding professional networks beyond this circle is a good idea and relatively easily accomplished. Getting to know others working in the same area can happen through correspondence (sending an e-mail commenting on this or that aspect of a colleague's recent publication, and perhaps attaching a paper of your own on a related theme is a common way to make contact, for example). Beginning faculty and sometimes senior graduate students will find that a certain amount of network-building happens without such directed effort on an individual's part. For example, being asked by journal editors to review manuscripts or write a book review in your area of expertise, by funding agencies (such as the National Science Foundation) to review research proposals, by colleagues to contribute to a textbook or other teaching resources are other sorts of opportunities to practice collegiality. Accepting such invitations and then doing the work well and in a timely fashion is one way to be a collegial member of broader scholarly and professional communities.

Another way to develop professional contacts is through participating in your discipline's professional organizations and their meetings. National and international-level conferences can be quite daunting at first for graduate students and beginning faculty. It sometimes seems as though everyone else knows each other, and you might find yourself nervously scanning the conference for a familiar face in the crowd. Participating in regional conferences or other smaller focused meetings is a great way to make professional contacts, but larger meetings and their associated organizations can be very welcoming too, especially if you join a specialty group with a focus in your area of interest. Through such groups and their newsletters and electronic bulletins, there are many opportunities to organize or co-organize sessions at relevant conferences and to volunteer to serve in some capacity (say as newsletter editor or paper prize judge). As you contribute your energy, ideas, and time to the group, you will find yourself developing many kinds of professional relationships that may take you in diverse directions, including those that lead to research or writing collaborations (Boice 1991, 1992). In addition, many multidisciplinary organizations (such as the International Studies Association or the American Geophysical Union) offer the opportunity to become part of an intellectual community that stretches beyond a home discipline. Just as in the case of departments, the kinds of collegiality practiced at the level of the discipline and in multidisciplinary settings involve give-and-take. Professional organizations and associations provide the infrastructure for more senior graduate students and early career academics to serve their wider community and participate in shaping these communities as they build their own professional relationships.

Similar understandings can be applied to relationships with colleagues in other units of your own university. Early career faculty (especially those from underrepresented groups) may find themselves asked to serve on committees at the college or university level. Whether or not to accept these sorts of invitations is a matter of service and should be discussed with one's department head, but serving on such committees can be another way of meeting people from across the college or university and can be a valuable way to build collegial relationships beyond a department. Many graduate students and early career faculty find it extremely productive and enjoyable to be part of multidisciplinary units, programs, or centers in addition to their departmental home. Women's studies, international studies, environmental science, and many other study/research/teaching areas may be organized in such a way as to bring together interested faculty and students around an interdisciplinary focus. Participating in such groups may be intellectually stimulating and personally rewarding, offering many chances to build collegial relations with an array of scholars (a point emphasized in Chapter 14 "Working Across Disciplinary Boundaries"). However, early career faculty would be wise not to focus only on contributing to an interdisciplinary unit at the expense of building positive and healthy relationships with their tenure-granting department. On the other hand, a department can be greatly strengthened by having faculty who develop a range of appropriate linkages and networks with other units in the university (Association of American Geographers 2006). Again, it is a matter of communication and being sensitive to how departmental colleagues and students may be included in, or positively affected by, any extra-departmental networks and groups you choose to contribute to.

CONCLUSIONS

The American Association of University Professors (AAUP) became very concerned that the issue of collegiality was being added to the three usual criteria for granting tenure and promotion: research, teaching, and service. Collegiality, they observed, was tricky to define and had the potential to be used against people who somehow were subjectively judged not to fit in and, as such, could lead easily into discrimination. Such, in fact, has been the claim of several persons, notably women, who were denied tenure on the basis of collegiality (AAUP 1999; see also Weeks 1996; Lewin 2002; Caesar 2006). Certainly, we would not wish to appear to be advising graduate students and early career faculty that they should be collegial in ways that are only always congenial and conformist, and we don't think that being collegial means never feeling free to express dissent (see Fogg 2002). For us, being collegial means being engaged in the life of the groups, associations, and networks that we inhabit as academics—and they would be much the poorer without open and reasoned debate and dissent. Being constructive and building positive professional networks doesn't mean always fitting in to established ways of doing things and always agreeing with everything that is going on. Rather, by being honest, respectful, sensitive, and strategic in your interactions and communications with colleagues, as we outlined above, you can cultivate meaningful and truly collegial relationships with colleagues even when there are difficult problems or disagreements to be dealt with. We hope that this chapter, with its discussion of the many ways collegiality can be built by early academics, will render the topic less hazy and stimulate early career colleagues to generate their own strategies for success. We have prepared activities for this book's web site that can get you started in this regard.

References

Åkerlind, G., and K. Quinlan. 2001. Strengthening collegiality to enhance teaching, research, and scholarly practice: An untapped resource for faculty development. In *To improve the academy,* eds. D. Lieberman and C. Wehlburg, 19:306–21. Bolton: Anker.

American Association of University Professors (AAUP). 1999. On collegiality as a criterion for faculty evaluation. Statement. http://www.aaup.org/AAUP/pubsres/policydocs/collegiality.htm (last accessed 1 September 2007).

Association of American Geographers. 2006. Healthy Departments initiative website. http://www.aag.org/healthydepartments (last accessed 1 September 2007).

Bennett, J. B. 1998. *Collegial professionalism: The academy, individualism and the common good.* Phoenix, AZ: American Council on Education and The Oryx Press.

Boice, R. 1991. Quick starters: New faculty who succeed. In *Effective practices for improving teaching,* eds. M. Theall and J. Franklin, 111–22. San Francisco: Jossey-Bass.

———. 1992. *The new faculty member: Supporting and fostering professional development.* San Francisco: Jossey-Bass.

———. 2000. *Advice for new faculty members.* Needham Heights, MA: Allyn and Bacon.

Caesar, T. 2006. The spectre of collegiality. *Symploke* 13 (1/2):7–17.

Diamond, R. 2004. *Preparing for promotion, tenure, and annual review: A faculty guide.* Bolton: Anker.

Duderstadt, J. J. 2000. *A university for the 21st century.* Ann Arbor: University of Michigan Press.

Fogg, P. 2002. Do you have to be a nice person to win tenure? *The Chronicle of Higher Education,* 1 February. http://www2.ucsc.edu/title9-sh/graduate/niceperson.htm (last accessed 1 September 2007).

———. 2006. Young Ph.D.s say collegiality matters more than salary. *The Chronicle of Higher Education,* 29 September.

Hanson, S. 2000. Networking. *The Professional Geographer* 52(4):751–58.

Jarvis, D. 1992. Improving junior faculty scholarship. In *Developing new and junior faculty,* eds. M. D. Sorcinelli and A. Austin, 63–72. San Francisco: Jossey-Bass.

Kennedy, D. 1997. *Academic duty.* Cambridge, MA: Harvard University Press.

Lewin, T. 2002. 'Collegiality' as a tenure battleground. *New York Times,* 12 July. http://www.physics.utoronto.ca/~peet/ofinterest/CollegialityasTenureBattleground.htm (last accessed 1 September 2007).

Modern Language Association (MLA). 2006. Taskforce report on evaluating scholarship for tenure and promotion. http://www.mla.org/tenure_promotion (last accessed 1 September 2007).

Nellis, M. D. 2006. Effective leadership. Coffman Leadership Institute keynote address. Manhattan, KS: Kansas State University.

Olsen, D., and M. D. Sorcinelli. 1992. The pretenure years: A longitudinal perspective. In *Developing new and junior faculty*, eds. M. D. Sorcinelli and A. Austin, 15–25. San Francisco: Jossey-Bass.

Phelps, P. 2004. Collegiality lessons. *The Chronicle of Higher Education*, 24 July. http://chronicle.com/jobs/2004/07/2004072701c.htm (last accessed 1 September 2007).

Silverman, F. 2004. *Collegiality and service for tenure and beyond: Acquiring a reputation as a team player*. Westport, CT: Praeger.

Sorcinelli, M. D. 1992. New and junior faculty stress: Research and responses. In *Developing new and junior faculty*, eds. M. D. Sorcinelli and A. Austin, 27–37. San Francisco: Jossey-Bass.

Stimpson, C. R. 2006. A dean's view of the MLA report. http://www.insiderhighered.com/views/2007/02/06/stimpson (last accessed 1 September 2007).

Weeks, K. M. 1996. Collegiality and the quarrelsome professor. *Lex Collegii* 20 (1), summer. http://www.collegelegal.com/lccolleg.htm (last accessed 1 September 2007).

Wheatley, M. 1992. *Leadership and the new science: Learning about organizations from an orderly university*. San Francisco: Barrett-Koehler.

CHAPTER 4

Balancing Personal and Professional Lives

Beth Schlemper and Antoinette M.G.A. WinklerPrins

The work of academic professionals is at once all-consuming and flexible. There is always more to read, more to research, more to teach, and more of many other endeavors. At the same time there is a tremendous amount of flexibility in regards to when we perform this work as it is not confined entirely by time or space. This flexibility is partly why overall satisfaction with academic employment is high, with 90 percent of academics reporting being satisfied with their work in one national study (Goldsmith, Komlos, and Schine Gold 2001).

Despite the fixed responsibilities such as teaching courses, office hours, and various meetings, faculty members have flexibility as to where and when they carry out many facets of their professional lives. This results in both a challenge and an opportunity to craft a life that is professionally rewarding and personally fulfilling. As an aspiring academic, learning to navigate the decisions you need to make regarding balancing your career and your personal life is key to your future success and life satisfaction. This chapter is a guide to the types of things to think about as you set your priorities.

In reality, there is no ideal balance and the concept is one that shifts over time as professional and personal demands change. It also involves other people in your life and the nature of your place of employment. This is not a zero sum game—it is not a case of choosing career *or* family, career *or* life's other interests. In fact, at times the boundaries between various aspects of our lives intersect and can be mutually enhancing (Ostriker 1998). Many successful academics have crafted lives that have brought them professional success and personal fulfillment. Although it is not easy, it is attainable through careful mindfulness and consideration. Active effort through self-assessment can be very helpful in this regard (see the activities in Chapter 2 on Career Planning by Janice Monk and Christine Jocoy).

Demographic change in the professoriate has prompted many academic institutions to re-examine their policies for supporting new faculty. Women now participate in the academic workforce in large numbers, and they can be found increasingly in the social and environmental sciences. More male aspiring academics are likely to be partnered with individuals with career ambitions, changing the way they go about their own careers (Miller and Hollenshead 2005). In the academic profession we also find a range of partnerships with different demands and expectations. "'Family' now means something more than just a married man and woman and their biological offspring. Many universities have responded to changes in the definition of family by broadening eligibility guidelines for inclusion in employee benefits" (Quinn, Edwards Lange, and Olswang 2004, 25).

Although having and raising children remains a goal and a choice that many aspiring academics pursue, the "life" part of the balance is also much broader than having children. Life can be filled with many activities that compete for the academic's time, some of which include the following:

- hobbies, including sports
- community involvement
- political activism
- commuting
- relationships
- separation or divorce from long-term partner
- acute illnesses of family/friends—both nearby and at a distance
- long-term eldercare or disabled-child care
- emergencies—nearby and at a distance
- estate settling at a distance

Some aspects of life can be planned and predicted, but much cannot. For example, "family commitments are not always something you can plan. Your [partner] . . . , content to manage the home in the early years of your [relationship] . . . , may now decide to go back to work, leaving a newly enlarged share of child and house responsibilities to you. Your aged father may need to move into your home. Your obedient, well-behaved child may turn into a juvenile delinquent. You may find yourself going through a messy, stressful divorce. This is life!" (Goldsmith, Komlos, and Schine Gold 2001, 157–58). Considering these and other potential situations that may alter our work/life balance, we all need to be able to accommodate shocks and uncertainty and situate ourselves so we can endure planned and semi-planned along with unplanned situations. Fortunately, many academic institutions are adopting policies that can help.

UNIVERSITY POLICIES FOR WORK/LIFE BALANCE

The American Association of University Professors (AAUP) published an entire issue of its flagship journal on the topic of balancing personal and professional lives (AAUP 2004), with one article dedicated to policy issues (Quinn, Edwards Lange, and Olswang 2004). Slowly, institutions are adopting family-friendly policies as a means of attracting and retaining professors. Faculty must take advantage of these policies because those that are never used may eventually disappear. By invoking what they are entitled to, faculty can slowly create an academic culture that will continue to attract quality, well-rounded people who can more easily balance their professional and personal goals.

Because of the length of schooling of most aspiring academics, there can be an unfortunate congruence of two difficult and stressful periods for many women and their partners: the early career stage and raising children. "Researchers concur that the model academic career path under the tenure system often conflicts with a faculty member's family responsibilities. Women continue to perform most care giving tasks in most U.S. families and are thus disproportionately affected by conflicts between the ideal academic career trajectory and family needs" (Sullivan, Hollenshead, and Smith 2004, 17). In light of this situation, many academics struggle with the decision of whether to put off having children until after the dissertation is completed, a permanent position is secured, or tenure is gained. As with many decisions, there is no easy answer. For some, attaining tenure may bring the permanence and sense of security they need to start a family. But for others, waiting increases the risk of never having children (or having fewer) as fertility declines and medical issues surrounding childbearing increase with age. Activity 4.1 is designed to help you understand the nature and complexities of these dilemmas for different people.

Research universities have been more likely than other academic institutions to have "family-friendly" policies (Ward and Wolf-Wendel 2004), but this is changing as all types of colleges and universities begin implementing work–family policies aimed at helping their faculty establish successful careers (Miller and Hollenshead 2005, also see the additional resources listed at the end of this chapter). You should investigate the family policies of any institution you are considering working for—in fact, we designed Activity 4.2 for this very purpose. Look for things such as

- the institution's commitment to assist with partner/spousal hire;
- maternity/paternity leave policies that go beyond the medical leave policy (sick days) and vacation, including the possibility of stopping your tenure clock;

- a family leave policy that lets you take a leave of absence to take care of aging or ill family members;
- university day care facilities and school-break and summer camp opportunities.

Knowing your goals and priorities as well as the various types of institutional policies related to family issues will enable you to make a more informed decision about your career-life options. Another way people are coping with their various professional and personal obligations is by forming support groups similar to the "Babies and Dissertations" group at the University of Wisconsin-Madison which formed through the efforts of a graduate student who used a listserv provided by UW's "Parent-Resource Office." This is a case of collective action, wherein the effort to find support came from those needing it instead of being initiated at the institutional level. The group was small, less than ten participants, and met every other week to offer a listening ear to academic and personal issues. Such groups work very well when run by highly motivated members, but often falter if those members move on.

FINDING THE "RIGHT" JOB

One of the most difficult issues facing aspiring academics is finding a job that is commensurate with their newly acquired skills and desired lifestyle. Keep in mind that, in general, liberal arts colleges demand a very high commitment to teaching and engagement with undergraduates, whereas faculty working at doctorate-granting universities tend to concentrate more heavily on research, grant writing, and advising graduate students. Overall demands and total working hours for faculty between these types of institutions actually do not differ much, but the feel of the demand may be different, and consequently so will the balance of work/life (Jacobs 2004).

Although many aspiring academics envision themselves becoming a college or university professor, the nature of these positions can vary. There are also many other types of career paths that can accommodate academic work. Consider the following range of possible appointments you may find yourself in someday (and which carry different sets of implications for achieving work/life balance):

- the tenure-track appointment (where the appointment progresses from assistant to associate with tenure to full professor);
- postdoc positions, often one to three years, usually focused on research but which may involve some teaching. Some academics do a series of these as they try to "break in" to the regular job market;
- full- or part-time instructor, adjunct or visiting professor positions (these generally do not lead to tenure but can sometimes include promotion from assistant to associate, etc.);
- research scientist (often with a hierarchy as well) at universities or in government agencies;

- employee with a federal, state, or local government agency;
- work in a nonprofit or nongovernmental organization;
- independent scholar/consultant;
- employment in private industry;
- some combination of the above positions, possibly part-time at several institutions.

Some feel that getting a job after graduate school is simply a matter of luck, but in actuality there is more to it than that. While the reality of the job market is that your first job may not be your first choice, it is possible to target a certain type of institution with a particular set of work demands and try to obtain a job there first. In graduate school, imagine the types of careers that you see yourself in the future, and try to get to know people who are already employed in those types of institutions. They may be able to provide you with some important insights about their jobs. Some institutions offer "preparing future faculty" programs to provide a taste of what the faculty life is like in a variety of academic institutional contexts.

In the next section, we highlight some of the many strategies available that can help you attain a desired work/life balance. Refer back to these strategies as you analyze in Activity 4.3 some of the issues that can potentially tip that balance unfavorably.

THE BALANCING ACT

Effective and efficient work.   As an aspiring academic it is imperative that you learn to be effective and efficient at time management (Wankat 2002). As Ken Foote recommends in Chapter 1, try to schedule your work for the time of day and in a place that permits you to be most productive. If you think more clearly in the morning then perhaps you can go in to work early while your partner handles the kids or work before the kids awaken. Think ahead about who stays home in the event of a sickness or a snow day to avoid misunderstandings and arguments. Always carry work home with you in case you cannot make it into the office (laptops and flash-memory cards can help in this regard). The key is being proactive so that you are prepared and can use your time well.

How to be a good colleague and still have a life.   In order to be a successful academician, you must build up some political capital in your department and institution. "Capital" is the give-and-take that comes with sharing the objectives of a larger goal such as the successful functioning of a department. This comes through service and requires a certain level of engagement with the daily business of a department. As Duane Nellis and Susan Roberts note in Chapter 3 "Developing Collegial Relationships in a Department and a Discipline," being a good colleague or departmental citizen contributes to a productive and positive working environment for everyone. Let us take a moment now to illustrate some of the ways that balancing personal and professional lives can intersect with collegiality.

Suppose, for instance, that your faculty meetings are scheduled for Monday afternoons from 3:00 to 5:00 p.m. every other week, and that this time happens to coincide with when your child needs to be picked up at the day care facility. Rather than always leaving the meeting early, on some weeks you may be able to ask a partner, friend, family member, or babysitter to pick up your child. If you must miss a meeting, be sure to follow up with colleagues and ask what transpired, and offer to help others.

Another example relates to teaching schedules. Often, the days and times of your course meetings are something that can be negotiated if done early enough and in a reasonable manner. Usually the chair of the department schedules classes and will schedule you wherever he or she sees fit *unless* you have indicated that you have preferences or times that are particularly difficult. Here we urge you to participate in some give-and-take, such as volunteering for the 8:00 a.m. general education class one semester in turn for an ideal schedule the next semester. Being a good colleague helps with the work/life balance. If you give when you are able, then when you need an accommodation, you can take.

Conferences.   Part of any academic career is the need to attend professional meetings regularly. This requirement can stress the work/life balance, but it can also enhance both dimensions. If you have personal obligations that could hinder travel to professional meetings, thinking strategically about meetings a year or so ahead could alleviate some of the stress. For example, if you both work in the same field, who will go to your shared meeting if you also have personal obligations? If you work in separate fields, it is possible that your meetings may be scheduled during the same time? If so, how will you deal with this balancing act? For partners in the same field, who need to attend the same professional meeting and have children at home, some specific strategies include the following:

- alternate who attends the primary conference this year versus next year;
- bring the kids and strategically plan the meeting regarding who stays with them and when;
- ask a relative or friend to take care of the children at home while you are gone;
- ask a relative or friend to come to the conference with you to take care of the children while you attend sessions.

For single parents or academics caring for aging relatives, balancing personal obligations with professional meetings can be quite tricky, but it is still possible. Some professional organizations, for example, offer on-site child care during a conference.

Despite well-intentioned efforts to prepare for professional journeys, there is always the potential for minor disasters upon your return that will steal time and attention away from work. Perhaps it isn't the best idea to leave several pounds of fish in the freezer if you plan to be away for six months (and the electricity is cut or the freezer breaks). Some of the practical matters that you would be wise to address before leaving for a conference or fieldwork include

- arranging for the mail and newspaper to be held or picked up for you;
- finding someone to take care of your plants and animals;
- obtaining the right amount of your medical prescriptions for the length of the trip;
- figuring out how you will pay your bills;
- hiring someone to clean your house if you can afford to do so;
- finding someone to house-sit during extended absences;
- leaving your contact information for appropriate friends, family members, or neighbors;
- having a neighbor or landlord observe your house for any problems (it is sometimes possible, depending upon where you live, to alert the police department of your absence so they will increase patrol on your street while you are gone).

These may seem like minor issues to consider, but they can save you from a major headache later.

Fieldwork. The expectation to conduct fieldwork is part of many social and environmental scientists' work lives. Fieldwork is time away from home, and this can lead to stress. Another aspect of conducting fieldwork is the potential of exposing yourself to dangerous situations. Individuals have different needs and feelings about being away, and this can change over time. The point is to think through what can be tolerated. Discuss, negotiate, compromise, and combine. As a scholar you may need to give up months in the field and learn to delegate research to students and assistants or to use different methods while your children are small. Your willingness to engage in fieldwork may change over time as the professional and personal aspects of your life change. Ultimately, fieldwork can be very productive from a scholarly perspective because it is often one of the few times you can completely immerse yourself in your work when you normally have family demands. But for this to be productive time you need to let go of managing the household while you are gone. Checking in periodically is recommended, but day-to-day decisions need to be left to the person you have entrusted with the task. Here are some specific ways to think about fieldwork:

- bring your children to the field (Figure 4.1)—this works well if you can accommodate their needs as well as your own in a reasonable manner and the area you work in is safe;
- arrange for parents, in-laws, or other trusted relative or friend to tend to the children while you are away if you do not have a partner or this partner is unavailable;

FIGURE 4.1 An academic father doing fieldwork in Turkey with his son.

Source: Kyle Evered, 2006.

- work out how long you can be away from home and/or how long
  your partner/children can do without you, then accommodate your
  research to those time constraints. This is potentially a very difficult
  issue, and it may take years to balance this.

The journal *The Geographical Review* has devoted two entire issues to stories
of doing fieldwork and its many facets, and we urge you to read sections of
it for ideas and approaches (DeLyser and Starrs 2001).

Health issues.   Another aspect of work/life balance is health. Life can get
very complicated with serious illness involving yourself, partners, parents,
children, and other significant people. The health of your family and friends
can also impact your productivity, as care of others requires time and disen-
gagement from academic pressure. This care–work, as Vicky Lawson puts it,
is important for our continued development as individuals and as scholars
and keeps us responsible and engaged in our communities that ultimately
enrich the discipline (Lawson 2004). Here too it is important to build up collegial relations with colleagues. Volunteer to help out when someone faces a
health crisis and needs course coverage—they, in turn, may help you if you
encounter difficulties of your own.

In order to remain a productive academic, you need to take care of yourself and have enough buffer and resilience to deal with health crises as these rarely occur at convenient times. High levels of stress are not healthful. Walking or bicycling to work or scheduling breaks at athletic facilities on or near the campus can be a healthy way to recharge and reduce stress.

The rest of life.   Academic work tends to be all consuming, but not all practitioners wish to conflate their life with their work. Some want to remain politically active and allocate time to those activities. Others choose to pursue the learning of another language or a musical instrument. But in the course of a career there may be better times for this than others. For example, waiting till beyond tenure to start a new language (unless, of course, this relates directly to your career advancement) or to learn piano may be better than trying to take on a new skill while also trying to make tenure.

Several academics have recounted their own experiences in balancing their professional and personal lives in autobiographical accounts (see, for example, Coiner and George 1998; Tuan 1999; Moss 2001; Gould and Pitts 2002). We encourage you to study their stories for additional advice that can help you balance all of your personal and professional responsibilities.

Additional Resources

American Association of University Professors (AAUP) Work and Family Resources (including Statement of Principles on Family Responsibilities and Academic Work and other work/life links): http://www.aaup.org/AAUP/issuesed/WF/ (last accessed 25 July 2007).

American Council on Education (ACE): www.acenet.edu (last accessed 25 July 2007).

The Chronicle of Higher Education: http://chronicle.com/ (last accessed 25 July 2007).

The College and University Work/Family Association (CUWFA), founded in 1994, is interested in issues related to balancing professional and personal lives: http://www.cuwfa.org/mc/page.do (last accessed 23 July 2007).

Drago, R., and C. Colbeck. *Final report from the mapping project: Exploring the terrain of U.S. colleges and universities for faculty and families to the Alfred P. Sloan Foundation*, 31 December 2003: http://lsir.la.psu.edu/workfam/MAPexecsummary.doc (last accessed 24 July 2007).

Gender Equity Project, Hunter College: http://www.hunter.cuny.edu/genderequity/ (last accessed 24 July 2007).

IUPUI Human Resources Department: http://www.hra.iupui.edu/worklife/ (last accessed 23 July 2007).

Massachusetts Institute of Technology, examples of policies implemented for work/life balance): http://web.mit.edu/fnl/women/women.html (last accessed 24 July 2007).

New York University's Office of Work-Life Services: www.nyu.edu/hr/worklife (last accessed 24 July 2007).

NSF/ NIH Survey of Doctorate Recipients, 1973 to the present (sponsored by the National Science Foundation and the National Institutes of Health): http://www.nsf.gov/statistics/srvydoctoratework/ (last accessed 24 July 2007).

Sullivan, E., C. Hollenshead, and G. Smith. 2004. Developing and implementing work-family policies for faculty. *Academe* 90(6): 16–19.

The Alfred P. Sloan Foundation (major contributor to work/life balance research and initiatives): http://www.sloan.org/main.shtml (last accessed 24 July 2007).

The Ohio State University, the Women's Place (information and support group): http://womensplace.osu.edu/ (last accessed 24 July 2007).

University of Arizona's "Life and Work Connections" Office: http://lifework.arizona.edu/ (last accessed 24 July 2007).

The University of Michigan's Center for the Education of Women (CEW) http://www.cew.umich.edu/ (last accessed 20 July 2007).

University of Washington, Academic Human Resources web site (follow Work and Life links): http://www.washington.edu/admin/hr/benefits/index.html (last accessed 24 July 2007).

References

AAUP. 2004. *Academe Online*. Entire issue on "Balancing Faculty Careers and Family Work." November–December 90 (6):3–31. http://www.aaup.org/AAUP/pubsres/academe/2004/ND/ (last accessed 23 July 2007).

Coiner, C., and D. H. George, eds. 1998. *The family track: Keeping your faculties while you mentor, nurture, teach, and serve.* Urbana, IL and Chicago: University of Illinois Press.

DeLyser, D., and P. F. Starrs. 2001. Doing fieldwork. *The Geographical Review* 91 (1–2):1–508.

Goldsmith, J. A., J. Komlos, and P. Schine Gold. 2001. *The Chicago guide to your academic career.* Chicago and London: The University of Chicago.

Gould, P., and F. R. Pitts, eds. 2002. *Geographical voices: Fourteen autobiographical essays.* Syracuse, NY: Syracuse University Press.

Jacobs, J. A. 2004. The faculty time divide. *Sociological Forum* 19 (1):3–27.

Lawson, V. 2004. Caring geography. *AAG Newsletter* 39 (10):3.

Miller, J. E., and C. Hollenshead. 2005. Gender, family, and flexibility—why they're important in the academic workplace. *Change* November/December:58–62.

Moss, P., ed. 2001. *Placing autobiography in geography.* Syracuse, NY: Syracuse University Press.

Ostriker, A. 1998. The maternal mind. In *The family track: Keeping your faculties while you mentor, nurture, teach, and serve,* eds. C. Coiner and D. H. George, 3–6. Urbana, IL and Chicago: University of Illinois Press.

Quinn, K., S. Edwards Lange, and S. G. Olswang. 2004. Family-friendly policies and the research university. *Academe* 90 (6):24–26.

Tuan, Y.-F. 1999. *Who am I? An autobiography of emotion, mind, and spirit.* Madison: The University of Wisconsin Press.

Wankat, P. C. 2002. *The effective, efficient professor: Teaching, scholarship, and service.* Boston: Allyn & Bacon.

Ward, K., and L. Wolf-Wendel. 2004. Fear factor: How safe is it to make time for family? *Academe* 90 (6):20–23.

CHAPTER 5

Succeeding at Tenure and Beyond

Susan M. Roberts

Completing a Ph.D. typically entails spending long periods of time focused on research and writing, often in relative isolation. Beginning assistant professors are sometimes surprised to find that this experience is unlikely to be repeated. Instead of working individually, a beginning faculty member finds himself or herself working as a member of a team (a department, for example); instead of having one major focus (the dissertation), there are now many competing expectations and demands on his or her time (teaching, advising, publishing, getting grants, and so on); instead of receiving advice and assistance, he or she is now advising and assisting others with their projects (be it through peer review of manuscripts or proposals or advising students). Needless to say, then, while the transition from doctoral student to faculty member is eagerly anticipated, it is often quite a disorienting experience.

For some, especially those from groups that remain underrepresented in academe, alienation and isolation can compound the situation. Being the only person of color in a department, for example, brings with it additional pressures that most other beginning academics do not bear. For everyone, moving to a new place and starting a new job can be stressful, but for those who find themselves the only gay person or the only woman in a department, for example, it is perhaps even more important to develop support networks within the institution and beyond it. While most universities and colleges have programs to assist minority students, faculty from underrepresented groups typically find that it is up to them to develop their own networks. Even where the climate might not be chilly for tenure-track faculty from underrepresented groups, research shows that the challenges they face are not well recognized or understood by colleagues and administrators and that change at the institutional and disciplinary level is very slow. The discipline of geography, for example, has a professoriate in the U.S. that is still overwhelmingly white and predominantly male (Toth 1997; Cooper and Stevens 2002; Sotello Viernes Turner 2002; Bonner 2003, 2004; Mahtani 2004, 2006; Monk, Fortuijn, and Raleigh 2004; Kobayashi 2006).

A range of professional development initiatives has been developed over the past twenty plus years in U.S. universities and colleges, and there are now many studies and advice manuals aimed at making the challenges entailed in this transition more transparent and at assisting aspiring academics to acquire the knowledge and skills that can be used to enhance their professional performance (e.g., Boice 1992, 2000; Lucas and Murry 2002). This chapter draws on this material to provide some background on tenure and some practical advice for those seeking tenure. Understandably, the tenure process can be a key locus of worry for beginning faculty members, but it is important to remember that tenure is a means, not an end. If a person truly enjoys doing research, writing, and teaching, then tenure is simply what will allow him or her to keep doing the work he or she loves. Moreover, tenure guarantees each faculty member the freedom to teach and do research in the ways that each one chooses, no matter if the topics and methods are unusual or unfashionable.

Tenure is not universal in higher education. Many countries in which Ph.D. graduates may obtain academic positions don't have tenure at all, or have a different form of tenure from that which prevails in the U.S. Having said that, this chapter primarily focuses on the U.S. Tenure is sometimes depicted as being primarily a matter of job security, but the security that tenure brings is a by-product of the goal of academic freedom. Tenure's primary function is to guarantee academic freedom by protecting faculty from the pressures to produce only certain types of research or scholarship or to teach only certain things in certain ways. Tenure is institutionalized protection for originality and creativity, the sparks that drive academia and grant it wider social value. It also works to strengthen faculty governance since tenured faculty are more likely to feel able to express views contrary to those prevailing in the administration (see Stimpson 2000). Being tenured, however, doesn't mean a person can't be dismissed for causes unrelated to academic freedom.

Tenure has always had its critics, and recently, in the mid-1990s, it faced some serious challenges from attempts to weaken it at several universities, including the University of Minnesota. Nonetheless, tenure has endured, although the proportion of faculty who are tenured or on a tenure track has fallen as universities and colleges increasingly turn to temporary faculty to fill teaching needs (Nelson and Watts 1999).

GETTING A (TENURE-TRACK) JOB

Doctoral students seeking careers in academe will likely enter a very tough job market, although there are pockets of geography and related disciplines where the demand is greater than the supply of new Ph.D.s. In general, though, graduate students need to prepare carefully and work hard to ensure that they are competitive on the job market. This means deciding early on in a doctoral program what kinds of jobs to aim for and ascertaining what kinds of experiences and qualifications are necessary to have a chance of getting such jobs.

In the U.S., there are a great many different types of institutions of higher education, and each has distinct expectations of its faculty members. Most colleges and universities in the U.S. are not like the research universities where doctorates are earned. Table 5.1, based on data collected by the Carnegie Foundation for the Advancement of Teaching, shows that the majority of higher education institutions in the U.S. are not "doctorate-granting" research universities and that the majority of students in higher education are enrolled in other types of institutions. As detailed on the Carnegie Foundation's web site (www.carnegiefoundation.org), each category identified in Table 5.1 contains a wide range of institutions: those with very large enrollments and those that serve very small numbers of students; those with multiple campuses and those with a single location; those that serve predominantly urban populations and those focused on rural areas; and publicly funded institutions as well as those that are private, which in turn may

TABLE 5.1   Types of U.S. colleges and universities.

Carnegie category	Number of institutions	Enrollment	Example(s)
Associates colleges	1,811	6,776,288	Chippewa Valley Technical College, WI; Oxnard College, CA
Doctorate-granting universities	282	4,907,275	University of Iowa, IA; Rice University, TX
Masters colleges and universities	665	3,905,461	SUNY at Cortland, NY; Rollins College, FL
Baccalaureate colleges	765	1,386,792	California State University-Monterey Bay, CA; Bucknell College, PA
Special focus institutions	807	571,493	Rose-Hulman Institute of Technology, IN; Union Theological Seminary, NY
Tribal colleges	32	17,599	Leech Lake Tribal College, Cass Lake, MN
Others	26	3,698	
Total	4,388	17,568,606	

Source: 2005 Carnegie Foundation Basic Classification.

be nonprofit or for-profit. A beginning faculty member is very likely to find himself or herself working in an institution that is quite different from any he or she has attended.

Learning about the many different types of institutions of higher learning in the U.S. is a first step to determining the setting that is likely to be the best fit for a prospective faculty member. In addition to the Carnegie Foundation's materials, the *Preparing Future Faculty* program's resources can be very helpful in this regard (see www.preparing-faculty.org). A person targeting jobs in a liberal arts type of college, for example, would strive to build up the kinds of experiences that would be useful in such a setting. In this case, experience teaching or working in interdisciplinary programs or leading student field trips might be assets. Building up a varied teaching portfolio takes more than a semester and, of course, can't be done at the last minute. To take just another example, if the goal is a tenure-track job in a major research university, having the Ph.D. in hand (or the dissertation almost done, perhaps with a defense date set) and evidence of success as a researcher in the form of publications is necessary. Because it can take a very long time (six months to a year is not uncommon) from submitting a manuscript to getting it accepted for publication in a journal, doctoral students who wish to list publications on a curriculum vitae should not wait until the last year of their degree to submit their manuscripts.

The positions open to new Ph.D.s are typically either of the visiting/adjunct/instructor type or of the tenure-track type. The former are proliferating in U.S. institutions and they offer fixed term contracts (from a year to five years) and tend not to come with the range of benefits and support available to a tenure-track appointee (Nelson and Watts 1999). For these reasons, many job seekers aim for a tenure-track position. However, the fixed-term position, while often not the desired first job, can be a very useful stepping stone into a tenure-track position. If you find yourself in a temporary or visiting position, although it will be a challenge, it is important to keep focused on building your CV in ways that will make you more attractive to those seeking a tenure-track colleague.

WORKING FOR SUCCESS AT TENURE TIME

Getting a tenure-track job is a big achievement but one that often brings with it a great deal of anxiety. Having accurate knowledge of the rules and procedures in effect at the institution is a first step to dealing with any worry caused by hazy ideas of how the tenure process may work. Each university or college has its own formal rules and regulations regarding faculty appointment, reappointment, review, promotion and tenure, and appeal. These are usually to be found in a university's governing rules or regulations and/or in any official "Faculty Handbook;" documents that increasingly are available on institutions' web sites. In addition to becoming knowledgeable about the rules and procedures, finding a more senior colleague or colleagues to act informally or formally as a mentor or mentors can be extremely helpful. A mentor does not have to be in the same department, but should be available to offer candid advice and guidance to a beginning faculty member.

A formal letter offering a tenure-track appointment will usually specify the "tenure home" for the appointee, that is, an academic unit (such as a department) in which tenure will be sought and granted. Typically, the person in charge of the specified academic unit (a department chair, for example) will have primary responsibility for administering all reviews of the appointee. In the case of a split appointment (with responsibilities in more than one unit), it is especially important to have in writing a clear statement of expectations and of how reviews will be conducted.

As soon as a tenure-track appointment has been accepted, the tenure clock starts ticking. In most institutions, progress to tenure will be formally reviewed after two or three years and the initial contract may be only for this first period, with reappointment depending upon a satisfactory review. It is crucial to make the first few years in a tenure-track job count, so that there is more than "potential" to be seen in your record at the time of any major pre-tenure review (Boice 1991). It is most useful to the tenure-track faculty member if these reviews are formal and entail written (rather than oral) reports, with specific (rather than general) comments on performance. Pre-tenure reviews can be helpful in pointing out areas where performance could be improved, and tenure-track faculty should take seriously any criticism, advice, or feedback they receive as a result of these reviews and adjust their strategies accordingly. Every college or university has formal procedures for handling any appeal made by a candidate who may wish to challenge an aspect of a review or a decision or who may wish certain matters to be reconsidered (see Baron 2003c for a case study).

Usually a person comes up for tenure in the first six or seven years in the job. It is often possible to come up early and it is sometimes possible for the tenure clock to be stopped for a period in the case of major illnesses or childbirth, for example, but whether the tenure clock can be stopped (and if so, under what circumstances) varies widely from institution to institution.

In most colleges and universities, tenure entails an elaborate review process involving assessments of the faculty member's performance by departmental colleagues and review committees at the college and university level (Baron 2003b). In some cases, such as at research universities, the tenure review will also include external reviews by leading scholars in the candidate's area of expertise. If this is the case, a candidate for tenure is invited to provide a list of suggested external reviewers, but in most cases some external reviewers will be people whom the candidate did not put forward (see Baron 2003a on external review letters). Candidates for tenure who have well-developed sets of professional networks will likely have a pool of people who are familiar with them and their work from which they can suggest names for consideration as external reviewers. Attending professional conferences and presenting your research are good strategies for building such networks.

All of these reviews are based on the candidate's tenure file or dossier—a formal compilation of evidence of performance. In addition to an up-to-date curriculum vitae (CV) and other materials relating to research, teaching, and service, the dossier contains documents prepared by the candidate in which he or she explains how his or her professional activities cohere and how his or her scholarship has been significant and developing over the years since the Ph.D. These personal statements are extremely important components of the dossier and will be carefully considered by each reviewer. There may be one single statement covering research, teaching, and service or separate statements for each of these areas (for example, a statement on teaching that would include a discussion of the candidate's teaching philosophy). Because many readers of the dossier will not be experts in the same field as the candidate, great care must be taken in crafting any such overview statements in ways that are understandable to nonspecialists. It is a good idea to have drafts of such materials read critically by peers and mentors before finalizing them. Suggestions about how to improve the overall style and approach are valuable, as are those of close readers who can double-check documents for typos and other errors that will potentially irritate reviewers.

Usually, the tenure dossier is organized in three areas: research, teaching, and service. The research category is often described in more general terms as "scholarly and/or creative achievement" to include the professional work of those in the arts, for example. In the sciences and social sciences, the label "research" is more common. The relative weight accorded to each of these three areas will vary from institution to institution, but they are nearly always all present as the major categories of assessment.

Research

There are many aspects to demonstrating success in research. Winning grants, presenting papers at conferences, and being asked to review others' research are all signs of research activity, but in most institutions the primary measure of a faculty member's research is publication. Increasingly, faculty members are expected to do research and have some publication record by the time they go up for tenure, even in those baccalaureate colleges where the emphasis is on teaching.

Straight answers to questions such as "How many articles do I need for tenure?" may be hard to obtain, so look instead at examples of recently tenured colleagues in your field (or something close) to see what was or was not deemed acceptable. Recently tenured colleagues may be willing to share their dossiers, and department chairs may have examples of tenure materials they could let a tenure-track person see. Even without access to such materials, new tenure-track faculty members should determine, as best they can, what sort of publications "count" for tenure. It might be the case that a chapter in an edited collection is regarded as less valuable than an article in a peer-reviewed journal, for instance. Thus, even if a person has been invited to contribute a paper to an exciting collection, he or she might be wiser to submit the paper to a journal. In some institutions, writing articles for the local press might be valued very highly, whereas in others it might not be. Publication strategies also vary by subdiscipline, with some characterized by more of a science model with multiauthored articles in peer-reviewed journals being the norm, whereas in others, perhaps closer to a humanities model, a single-authored book is the expectation.

Ideally, after finding out the formal requirements regarding research and publication, a new tenure-track faculty member should develop a plan for research and a related strategy for publication. Seeking feedback from the chair and/or other mentor(s) on your plan is also important. Should you write a book, for example, or does your institution value articles in peer-reviewed journals much more highly? If a book is regarded as a good strategy, find out what presses are regarded positively. A university press may be valued more than a commercial press, for example. Above all, remember that publications are the "coin of the realm" when it comes to tenure. It is rare that anything else can make up for a weak publication record. The old saying "publish or perish" is accurate even if it is hackneyed.

Publishing, though, is the end result of research and writing. So, these activities need to be prioritized: time needs to be made for them in the new faculty member's busy days and weeks. If a person's research entails fieldwork, equipment, or expensive data, grants will need to be applied for and won. Many colleges and universities offer assistance to faculty in researching funding options, preparing proposals and budgets, and submitting grant applications, and such services can be invaluable. Research on the strategies of successful faculty members shows that it is wise to make time to write and protect this time as much as possible from encroachment. Successful faculty members don't delay submitting a manuscript even though they may know this or that aspect of it isn't perfect (Boice 1991). Time goes by quickly while the tenure clock ticks relentlessly and, as mentioned above, the time from submitting a manuscript to its appearance in print can be long. It is important not to be fooled into thinking you have plenty of time before the tenure review; get an early and fast start to building up your publication record in ways that are appropriate for your institutional setting.

Teaching

Teaching responsibilities vary widely from institution to institution. At the majority of institutions faculty members typically teach four courses per semester. These might be four different courses or sections of the same course or some mixture of the two. In some institutions the teaching load is lighter, such as at doctorate-granting universities where the norm is for faculty to teach two courses per semester. In all cases, there is usually some consideration given to how many different courses a person teaches, sometimes referred to as how many preps he or she is responsible for in total or in a given period.

It is quite likely that a tenure-track faculty member will be expected to contribute to teaching at all levels from the first year courses or general education courses, through to courses for majors and/or honors students and, where relevant, for graduate students. The average number of students in a course can vary greatly too. A person could teach 4-4 (four courses per semester) but have fewer students than someone teaching 2-2. Most faculty members teach courses that are not directly in their area of specialization as a contribution toward a department's curriculum or toward fulfilling an institution's specific teaching mission. Being flexible in the courses you are willing to teach (or develop) is usually regarded positively by colleagues. Even when teaching courses that are not directly related to your research specialty, there are many potential ways to make connections between teaching and research (such as by inviting a colleague to share relevant data or slides from his or her fieldwork).

Regardless of the institution, it is important that new faculty members guard against being given an unfair teaching and advising load. This can be a particular problem for faculty with joint appointments or split teaching responsibilities (for example, half in geography, half in African-American studies) or even for those who prove themselves to be highly competent instructors or advisors. Negotiating assignments might not be possible, but discussing the matter with a mentor may be helpful to a faculty member who is seeking to raise the issue constructively. Mentors can be helpful in thinking through how a conversation on the topic with the chair or other appropriate administrator could be managed.

It is important that each tenure-track faculty member build up a record of being an effective instructor. Students may not always be respectful of younger or early career faculty, especially when the faculty member is from an underrepresented group, but new faculty members should always set the tone by being respectful of students. Experimenting with teaching techniques becomes easier with experience, and most faculty find it very satisfying to build on their successes in this regard.

Student evaluations are typically important elements in faculty review and assessment (including tenure). Teaching evaluations can have their problems (see Sinclair and Kunda 2000), but they can also flag areas where improvement could be made. If this is the case, it is a good idea to discuss strategies with your chair and make sure he or she knows that you are attending to any problems. Boice's research (1991) shows that beginning career faculty can improve their teaching performance by inviting constructive feedback from colleagues and faculty development professionals on campus.

Service

Service expectations also vary widely from institution to institution. Nearly every institution will expect tenure-track faculty to have a demonstrable record of effort and accomplishment in service. In some cases, service may not be a central component of evaluation, whereas in an institution with a core commitment to urban outreach for example, a candidate for tenure might be expected to demonstrate a contribution to this mission. Despite a tendency by some to regard service only in terms of the opportunity costs it effects on other areas (teaching and research), it is a vital part of academic life (an issue that is discussed at length in Chapter 2 "Career Planning: Personal Goals and Professional Contexts" by Janice Monk and Christine Jocoy).

The quality of research and teaching depends on the service of those reviewing manuscripts for journals, running listservs, serving on administrative committees, organizing visiting speakers, and arranging campus and community events (to name but a few activities that may be counted as "service"). In a recent article on the value of service, geographer Susan Hanson points out that it is through service that our institutions and disciplines may be positively changed and shows how pioneering feminist geographers made meaningful change through their efforts in reshaping the discipline's organizations. Having said this, it is unwise to take on too much service before tenure. Faculty members from underrepresented groups often experience particular pressure in this area (Bonner 2006). Again, this is an area where a mentor can offer advice on what sorts of service opportunities to accept or decline.

POST-TENURE REVIEW

Once tenured, faculty members continue to be regularly assessed and reviewed in most institutions. Being considered for promotion (for example, to the rank of full professor) will entail a more formal review process, similar to that in place for tenure. Many U.S. colleges and universities have formalized post-tenure review procedures. Post-tenure reviews are either conducted periodically (every three years, for example) or only if triggered by some specified level of poor performance (Montell 2002). Faculty members who are assessed as performing poorly in some or all areas are usually offered some kind of assistance in improving their performance, although the threat of reassignment or even dismissal is usually present as well. Some are critical of post-tenure review as an assault on tenure or resent it simply because it adds another layer of review, assessment, and paperwork to faculty members' lives. But others view post-tenure review as a valuable system that aims to identify underperforming colleagues and help them get back on track.

CONCLUSIONS

To wrap up this chapter, I offer some direct advice:

1. Don't rely on gossip or hearsay about how the tenure process works at your institution. Get the facts straight and get them early on. Read your institution's official rules about tenure and review. Consult with your department chair or other appropriate administrator about tenure procedures and expectations. Go to any university-offered workshops that explain the promotion and tenure system.

2. Listen to senior colleagues and any colleagues who have recently successfully gone through the tenure review: even if they are not formally your mentors, they are often in positions to offer useful information and advice. Additionally, find at least one mentor, including someone who has succeeded professionally in a related field at your institution, but who is not directly responsible for reviewing your progress. Many colleges and universities have mentoring programs.

3. Be organized: keep your CV up to date and keep paper and electronic files where you can build as you go all the materials you will need to include in your reviews and in your tenure dossier. Under research, file copies of all published works; invitations to present your work or to participate in prestigious conferences or workshops; notices of any awards or honors; evidence of any further professional training you may have completed; research proposals submitted (whether successful or not); and any reviews of your published work. Under teaching, keep copies of all your syllabi; a record of what courses you have taught in each semester and how many students were in each; all teaching evaluations; any material indicating your positive impact on student learning, including any appreciative letters or notes from students; and records of any pedagogical workshops or the like that you may have attended. Under service, keep a detailed record of your contributions and any materials that may document your positive contribution in this area. Request and keep written copies of any reviews conducted of your performance.

4. Don't spin your wheels worrying about tenure. Channel any anxiety you may feel into productive work in areas directly related to strengthening your record. Find a peer group, or just a peer, with whom you can share your feelings and strategies for working productively. If you are a member of an underrepresented group, you will likely encounter situations that you might feel more comfortable discussing with peers in the same or similar situation, and perhaps being part of a formal or informal network at the national level could be a resource in this regard.

5. Work as productively as you can. Each person has different optimum work habits, and it is important to know yourself well enough to

recognize your own and to plan your daily, weekly, and yearly schedules accordingly. For example, if you are a morning person who has your most creative and productive writing time before dawn, endeavor to structure your days so you can concentrate and work well during that time. Likewise, keep yourself (at least somewhat) balanced by continuing any activities that offer you down time—whether it is running marathons or knitting sweaters.

It is the case that most tenure-track faculty worry at least a little about tenure, but it is important to remember that tenure is only worth having if you are happy being a faculty member and find research, teaching, and service to be enriching aspects of your life, not just some categories in a file drawer. With tenure you will be able to invest even more fully in your work, your students, and your institution, a step that should be fulfilling on a personal level even as it is a professional validation of your contributions.

References

Baron, D. 2002a. Getting promoted. *The Chronicle of Higher Education*, 5 September. http://chronicle.com/jobs/2002/09/2002090501c.htm.

———. 2002b. A look at the record. *The Chronicle of Higher Education*, 7 November. http://chronicle.com/jobs/2002/11/2002110701c.htm.

———. 2003a. External reviewers. *The Chronicle of Higher Education*, 17 January. Available at: http://chronicle.com/jobs/2003/01/2003010701c.htm.

———. 2003b. The tenure files: Getting through the college. *The Chronicle of Higher Education*, 14 February. http://chronicle.com/jobs/2003/02/2003021401c.htm.

———. 2003c. When tenure fails. *The Chronicle of Higher Education*, 10 June. http://chronicle.com/jobs/2003/06/2003061001c.htm.

Boice, R. 1991. Quick starters: New faculty who succeed. In *Effective practices for improving teaching*, eds. M. Theall and J. Franklin, 111–22. San Francisco: Jossey-Bass.

———. 1992. *The new faculty member: Supporting and fostering professional development*. San Francisco: Jossey-Bass.

———. 2000. *Advice for new faculty members: Nihil Nimus*. Boston: Allyn & Bacon.

Bonner, F. A. II. 2003. The temple of my unfamiliar: Faculty of color at predominantly white institutions. *Black Issues in Higher Education* 20 (18):55.

———. 2004. Black professors: On the track but out of the loop. *The Chronicle of Higher Education* Review, 11 June.

Cooper, J. E., and D. D. Stevens. 2002. *Tenure in the sacred grove: Issues and strategies for women and minority faculty*. Albany, NY: SUNY Press.

Hanson, S. 2007. Service as a subversive activity: On the centrality of service to an academic career. *Gender, Place and Culture* 14 (1):29–34.

Kobayashi, A. 2006. Why women of colour in geography? *Gender, Place and Culture* 13 (1):33–38.

Lucas, C. J., and J. W. Murry. 2002. *New faculty: A practical guide for academic beginners*. New York: Palgrave Macmillan.

Mahtani, M. 2004. Mapping gender and race in the academy: The experiences of women of color faculty and graduate students in Britain, the US and Canada. *Journal of Geography in Higher Education* 28 (1):450–61.

———. 2006. Challenging the ivory tower: Proposing anti-racist geographies within the academy. *Gender, Place and Culture* 13 (1):21–25.

Monk, J., J. D. Fortuijn, and C. Raleigh. 2004. The representation of women in academic geography: contexts, climate and curricula. *Journal of Geography in Higher Education* 28 (1):83–90.

Montell, G. 2002. The fall-out from post tenure review. *The Chronicle of Higher Education*, 17 October, Careers section.

Nelson, C., and S. Watts. 1999. *Academic keywords: A devil's dictionary for higher education.* New York: Routledge.

Richardson, J. T. 1999. Tenure in the new millennium: Still a valuable concept. *National Forum*, Vol. 79, No. 1.

Sinclair, L., and Z. Kunda. 2000. Motivated stereotyping of women: She's fine if she praised me but incompetent if she criticized me. *Personality and Social Psychology Bulletin* 25 (11):1329–42.

Sotello Viernes Turner, C. 2002. Women of color in academe: Living with multiple marginality. *The Journal of Higher Education* 73 (1):74–93.

Stimpson, C. A. 2000. A dean looks at tenure: An interview with Catherine Stimpson. *Academe* 86 (3).

Toth, E. 1997. *Ms. Mentor's impeccable advice for women in academia.* Philadelphia: University of Pennsylvania Press.

SECTION II

Developing and Enhancing Teaching and Advising Skills

The chapters in this section address key issues relating to learning, teaching, and advising. For most graduate students and early career faculty the roles of teacher and advisor cause more anxiety at the start of their career than any other issues but, in the longer term, serve as one of the most satisfying and enjoyable parts of academic life. The reason for the stress is that, although most graduate students gain some experience in the classroom as a teaching assistant or lecturer, it is not unusual for faculty to assume their first appointment without ever having independently developed or taught a course. Still fewer new faculty will have had any systematic introduction to course or curriculum design, learning theories, strategies for assessment and evaluation, or issues of ethics in learning and teaching. It is only later, through on-the-job experience—perhaps combined with some independent reading and study—that faculty gradually acquire the background necessary to thrive as teachers. As Ken Bain (2004) observed in *What the Best College Teachers Do*, many award-winning professors develop an intuitive understanding of some of these key pedagogical issues. But why wait? A little bit of early help can reduce stress and allow for more efficient use of scarce time. Early career faculty can then concentrate on creating significant learning experiences, rather than reinventing pedagogical wheels.

In this section we have chosen to highlight five topics that we see as among the most critical to getting a quick start on an academic career:

designing significant learning experiences and courses, strategies for active learning, student advising, ethical issues in teaching, teaching diverse students and teaching for inclusion. We could have picked others but, from the many Geography Faculty Development Alliance workshops we have led since 2002, these are the topics that have been consistently rated most important by participants. We could have devoted more space to these and other pedagogical issues, but chose not to. As we argued in our introduction, the issue of balance is fundamental to our vision of academic life. Although teaching may cause more stress initially, we feel it is important— from the start—to see teaching, research, service, outreach, and our personal lives as interconnected. We have selected the chapters in this section as good starting places for developing and enhancing teaching and advising skills; other excellent sources exist for readers interested in more detail and guidance.

For geographers and scholars in related fields, we have prepared another book, *Teaching College Geography*. Included in that book is a complete guide to getting started in the college classroom with additional chapters and materials on promoting the scholarship of teaching and learning, GIS and mapping tools for reasoning and critical thinking across the curriculum, developing significant learning in large classes, teaching in the field, and geography and global learning. For scholars and scientists in other fields we recommend McKeachie and Svinicki (2005), Fink (2003), Davis (1993), Biggs (1999), and Angelo and Cross (1993). Also of great value are Walvoord and Anderson (1998), Wiggins (1998), and Wiggins and McTighe (2005).

OVERVIEW OF THE CHAPTERS

Dee Fink and Melissa Ganus's chapter is one of the most important in this book. As the authors point out, "The one task that college professors are least prepared ... is how to design powerful learning experiences. This is why so many of our problems stem from this aspect of our teaching and will not go away until we learn how to design our courses more effectively." In the absence of good design principles, most early career faculty find themselves under pressure to teach well, but without the concepts and tools needed to do so. Instead, the sense of pressure seems to increase as they try to make do with improvised solutions. Dee has devoted his career to helping faculty overcome this problem. The chapter he has coauthored with Melissa presents the key arguments of his book *Creating Significant Learning Experiences* (2003). When used with the activities contained in the self-directed guide available from the website for *Aspiring Academics*, readers have a step-by-step guide to creating better learning experiences. Dee and Melissa are not suggesting a "quick fix" to the challenges of course design. Time and careful thought are needed to implement their student-centered, systematic approach to designing courses. What they are offering are principles for "smart" design—the theoretical and practical knowledge needed to get the most out of the time we invest in preparing classes.

In the second chapter, Eric Fournier takes up the issue of active learning. He provides an excellent theoretical rationale as well as practical advice about how to experiment with active learning strategies in the classroom. We highlight active learning in this book because it is fundamental to the vision of integrated course design presented by Fink and Ganus. For them, focusing on active learning is one of the four most important foundational steps in designing courses. It always involves providing learners with new information and ideas, an opportunity to do or observe something, and time to reflect on their learning. We stress this point because the active learning that Eric advocates is often quite different from some of the models of teaching—such as lecturing—with which we as instructors are most familiar and most comfortable, particularly when faced with time pressure. Eric provides clear suggestions about how to begin to think "actively" about learning and teaching, how active learning can be aligned with learning outcomes, and links to a range of web and print resources that can make it easier to develop active learning strategies for your classes.

The chapter by Fred Shelley and Adrienne Proffer focuses on student advising—one of the topics rarely discussed in graduate school. This situation seems to rest on a sort of *Catch-22* assumption that, since we have all had good advisors to succeed in our careers, we will all succeed as good advisors. But this need not be the case. It is important instead, as Shelley and Proffer point out, to see advising as a fundamental part of the learning experience. This is "developmental advising," which stresses the need to help students take charge of their learning, personal development, and career preparation. This view elevates advising from simply answering routine questions to a chance to engage students critically in issues relating to their academic work and to their careers beyond college. Without such perspectives, it is too easy to slip into comfortable patterns—to focus on advising students whose abilities, backgrounds, and aspirations are most like our own. Instead, good advising involves engaging *all* students, irrespective of their abilities, backgrounds, and aspirations, in dialogue about their academic performance and life plans.

This section continues with another topic not often included in graduate curricula—the ethical dimensions of teaching. But, as author James Ketchum points out, in the absence of an understanding of the issues involved, ethical issues can take us by surprise. Suddenly, we are faced with a case of plagiarism, an episode of cheating, or an instance of unprofessional conduct for which we lack the background to resolve fairly and ethically. In his chapter, he draws attention to the codes of behavior that support professional standards of competency, encourage fairness and respect for our colleagues and students, oblige us to teach in a manner consistent with institutional goals and values, and lead us toward a personal dedication for continued improvement and development in the area of teaching. He argues that we have responsibilities relating to the instructional content of our courses; our pedagogical competence; the valid and timely assessment of student work; the intellectual development of students; the confidentiality and privacy of our interactions; the teaching of sensitive topics; and the avoidance of exploitative, discriminatory, or harassing activity on our parts or by other students. It is hard to anticipate all of the possible ethical and moral scenarios that can arise in teaching, but James provides a starting place for analyzing some of the problems we are likely to confront as well as the concepts and resources that may be of most assistance.

Minelle Mahtani's chapter concludes this section and raises important issues about diversity and inclusion in college teaching. She pays particular attention to how we can create a culture of support that welcomes a diverse range of voices and identities both inside and outside the classroom. Minelle notes that our students and the worlds in which we live are increasingly diverse. One of our challenges as educators is helping students develop the tools they need to understand these changes—"Taking diversity seriously highlights the myriad assumptions one makes when developing a critical pedagogical practice, and it demands that one critically contemplates what teaching is for, ultimately." Minelle draws attention to research indicating how vital it is to understand, acknowledge, and support students who both possess and express diverse perspectives. She argues that graduate students and early career faculty can make positive change in their courses through more diverse pedagogical practices and curriculum design and by questioning the assumptions and preconceptions we and our students bring into learning environments. Although Minelle focuses largely on issues of inclusion and exclusion in teaching, it is clear that her points apply just as well to our advising, efforts at cultivating collegiality, and many other aspects of academic life. The change in the climate and culture of higher education that we advocated in the introduction to this book revolves fundamentally around making academic life more inclusive and welcoming to all.

References

Angelo, T. A., and K. P. Cross. 1993. *Classroom assessment techniques: A handbook for college teachers*. San Francisco: Jossey-Bass.

Bain, K. 2004. *What the best college teachers do*. Cambridge, MA: Harvard University Press.

Biggs, J. 1999. *Teaching for quality learning at university: What the student does*. Buckingham, U.K.: Open University Press.

Davis, B. G. 1993. *Tools for teaching*. San Francisco: Jossey-Bass.

Fink, L. D. 2003. *Creating significant learning experiences: An integrated approach to designing college courses*. San Francisco: Jossey-Bass.

McKeachie, W. J., and M. Svinicki. 2005. *McKeachie's teaching tips: Strategies, research, and theory for college and university teachers*, 12th ed. Boston: Houghton Mifflin Co.

Walvoord, B. E., and V. J. Anderson. 1998. *Effective grading: A tool for learning and assessment.* San Francisco: Jossey-Bass.

Wiggins, G. 1998. *Educative assessment: Designing assessments to inform and improve student performance.* San Francisco: Jossey-Bass.

Wiggins, G., and J. McTighe. 2005. *Understanding by design,* 2nd. ed. Upper Saddle River, NJ: Pearson Education, Inc.

Designing Significant Learning Experiences

Dee Fink and Melissa Ganus

Teachers affect eternity; they can never tell where their influence stops.

HENRY BROOKS ADAMS, AMERICAN HISTORIAN

WHY SHOULD YOU CARE ABOUT COURSE DESIGN?

Your course design has the potential to facilitate significant learning. When anyone teaches—good teacher, bad teacher, innovative teacher, traditional teacher—we all engage in four tasks that are fundamental to teaching, as illustrated in Figure 6.1. The first two tasks happen, for the most part, before a course begins:

- *Mastering knowledge of the subject matter:* We have to know something about the subject matter we are trying to teach.
- *Designing of learning experiences:* Collectively we make many decisions about how we want a course to unfold: Are we going to use a textbook or multiple readings; use small groups or not; use frequent writing activities; use one midterm and a final, or have weekly assessment activities?

FIGURE 6.1   Four tasks fundamental to teaching.

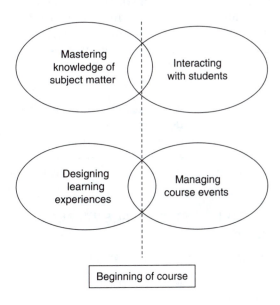

Once the course begins, the other two factors come into play and the most important one is the first:

- *Interacting with students:*   We interact with students in many ways: lecturing, leading discussions, visiting with students in office hours, dialoguing via e-mail, and so forth.
- *Managing the course:*   We must keep track of data and materials: who is enrolled and who has dropped; who took the exams and who needs to do a make-up exam; who turned in their paper and who did not, etc.

As a group, college professors are not equally well prepared for each of these four tasks. Almost all have lots of formal preparation for their knowledge of the subject matter they teach. Everything in graduate school is focused on enhancing this; faculty candidates are grilled on it, and tenure decisions depend on it. Most professors also handle the course management tasks in a satisfactory manner and have developed human interaction skills to a level that is sufficient for that part of teaching.

The one task that college professors are least prepared for and do not develop from ordinary life experiences is designing powerful learning experiences. So many of our problems as early career faculty stem from having little preparation in teaching and learning issues, and these difficulties will not be resolved until we learn how to design our courses more effectively.

In this chapter and the accompanying online activities, we strive to provide you with clear and easy-to-use course design methods and rationale. Whether you are a new or experienced teacher, we believe these materials will help you. Even if you only have a small amount of time, you can still design significant learning experiences by:

- understanding the benefits of using a learning-centered, integrated, systematic approach to designing your courses,
- developing the ability to use this twelve-step model for designing courses that facilitate significant learning for your students, and
- helping you teach others how they, too, can design courses this way.

WHY SHOULD YOU USE STUDENT-CENTERED COURSE DESIGN?

Current Focus of Our Course Design Procedures

Research shows that students learn poorly in traditional courses. For centuries, the most common way of designing courses has been to focus the course on a "list of topics" and then for the teacher to provide lectures, summarizing his or her understanding of each topic. This has the advantage of giving the teacher the feeling of being in control, but it also has several major disadvantages. First, unless the teacher is extraordinarily dynamic, students often get bored after a few weeks of continuous lecturing. Second, the student learning that results from courses designed this way has a short half-life—research indicates that one year later, students can recall only 15 percent of what they learned (Saunders 1980). Another disadvantage, most importantly, is that this way of teaching does very little to promote the kind of learning seen as valuable in today's society—critical thinking, the ability to read and write well, decision making, or working with others on a team to solve open-ended complex problems, especially with people different from oneself.

A second common way of designing courses is the "list of activities" approach. It is like the "list of topics," except that every week or two, the teacher inserts an activity into the sequence of lectures, for example, role playing, case studies, a simulation exercise, and so forth. This approach usually succeeds in making courses more interesting, and it appeals to a wider range of student-learning styles. But it is still like putting tasty icing on a bland cake: It tastes better but it does not make a big improvement in the quality of student learning.

New Paradigm for College Teaching

During the last fifteen years or so, a major paradigm shift has been occurring in the field of college teaching whose significance is akin to the Copernican theory of the solar system and the DNA model in biology. The article that initially articulated this shift called it "From Teaching to Learning—A New Paradigm for Undergraduate Education" (Barr and Tagg 1995). Basically this means that we, as college professors, need to be focused on learning rather than on teaching when we teach a course. Although this initially seems like a subtle difference, the implications are major. When we focus on teaching, we only need to worry about whether we "know our stuff" and can communicate in a coherent manner. When we focus on learning, the former requirements are clearly not sufficient; we also need to worry about whether students have learned something significant. The bottom line is not the quality of our teaching but the quality of student learning. If the learning is not good, the teaching is not good.

How Students Learn

Once we start paying more attention to student learning, we realize that we need a better understanding of how students learn. A number of books have been published for college teachers in the last decade or so that address this question (Bransford, Brown, and Cocking 1999; Zull 2002; Svinicki 2004). These and other books frequently refer to the constructivist view of how people learn. According to this view, it is not possible to "transmit" knowledge. You can transmit information, but then learners have to take that information and construct their own understanding, that is, knowledge, of what it means, as well as when, where and how to use that information. There is one important variation of this view called social constructivism that posits: When we construct our understanding of something, it usually does not work best to do this alone. We do it better when we dialogue with others about it. This is what underlies the growing awareness of the value of small group work, as well as individual and group reflection of learning.

When teachers shift their primary focus to the quantity and quality of student learning and when they realize that courses need to contain activities that enable students to construct their own knowledge, they find themselves faced with the need for a radically different way of putting their courses together. What they need is a set of course design procedures that:

- begin by asking what the important kinds of learning are that this course should promote,
- work systematically through other important decisions, for example, selecting appropriate learning activities and assessment activities,
- show how to integrate or align the key parts of the course.

In this chapter, we will introduce the model of Integrated Course Design which meets these criteria. In Activity 6.1 (available on this book's web site), you will find worksheets and other resources to help with the task of creating a high-impact, engaging course for your students—and one which will also be extraordinarily exciting for you to teach.

A DREAMING EXERCISE: WHAT IF YOU COULD TEACH THEM ANYTHING?

Now, we want to encourage you to pause your reading and interact with this material by completing the following reflection, designed to help you become more fully aware of some of your own values in relation to high-quality learning.

Imagine yourself preparing to teach an ideal course. You will be teaching in a perfect environment, where all your situational factors support whatever you want to do. You are a perfect teacher. Your administration gives you full discretion and support to teach whatever you want to teach, however you want to teach it, with unrestricted funding. Your students are ideal learners, motivated and engaged, making substantial progress on learning goals with every learning activity you provide, eager to do everything you ask of them, completing every reading and writing assignment, and exceeding all your grading criteria. These students can and will learn, remember, and use everything you want them to learn; the only limits are in what you can imagine them learning. *What would you most want them to learn?*

There are no wrong answers. Take a few minutes *now* to consider your initial responses to this question and write them down, if possible; the time you invest now will help your understanding and ability to apply the concepts presented in this chapter. What words or images come to mind?

With this exercise, you are getting at your core visions for your teaching potential. If, as you design courses for the imperfect real world, you continue to reflect on this question, you will see opportunities present themselves that allow you to teach at least some of what you believe is most important for your students to learn.

If you want to compare your dreams for student learning to the dreams of other professors, here are some typical responses:

"My dream is that students, one to two years after the course is over, will..."

- Apply and use what they learn in real-life situations.
- Find ways to make the world better, be able to "make a difference."
- Develop a deep curiosity.
- Engage in lifelong learning.
- Experience the "joy of learning."
- Take pride in what he or she has done and can accomplish, in whatever discipline or line of work he or she chooses.
- Think about problems and issues in integrated ways, rather than in separated and compartmentalized ways. Students will see connections between multiple perspectives.
- See the need for change in the world and be a change agent.
- Be creative problem solvers.
- Develop key life skills, for example, communication skills.
- Stay positive, despite the setbacks and challenges of life and work.

DIFFERENT MODELS OF THE COURSE DESIGN PROCESS

Now that you have identified what kinds of learning you would most want for your students, the question is, What can you do to make that kind of learning happen—for the majority of your students, the majority of the time—on a regular and intentional basis? The short answer to that question is, *You need to design that kind of learning into the course.*

There are multiple models for this process, some called instructional design, others called course design. Teachers in the K-12 realm of education have studied this process for decades and hence have numerous books and models. One of the more significant models developed for K-12 teachers is that of "Backward Design" created by Wiggins and McTighe (2005). Backward Design calls for teachers to identify their learning goals first, and then work backwards to create appropriate assessment procedures, and finally learning activities.

The earliest book on instructional design specifically for higher education in modern times was written by Robert Diamond (1989, 1998). His process is more complicated but is especially valuable in paying attention to the specifics of the learning situation and the particular characteristics of the subject being taught. More recently, John Biggs created a model for professors in higher education called "Constructive Alignment." It focuses on the same three factors highlighted in Backward Design.

In *Creating Significant Learning Experiences: An Integrated Approach to Designing College Courses*, Dee Fink (2003) created a model called "Integrated Course Design." Although the Backward Design, Constructive Alignment, and Integrated Course Design models were each created independently, they all highlight the importance of three central decisions:

- formulating important learning goals,
- selecting appropriate assessment and feedback procedures, and finally
- developing effective learning activities.

Integrated Course Design is somewhat different in that it also includes specific guidelines for knowing how to do each of these tasks well. The remainder of this chapter will describe the elements of the Integrated Course Design model.

HOW CAN YOU APPLY THE INTEGRATED COURSE DESIGN MODEL?

Work with the step-by-step process below—this process is concerned with all the decisions that we make—before the course begins—about what we want the course to accomplish and how we want it to unfold. The Integrated Course Design model is simply a framework for highlighting the key decisions in course design and providing a logical sequence in which to address them. The full model consists of twelve steps as follows:

Step 1   Identify important situational factors that could affect your course.

Step 2   Identify your top student-centered learning goals.

Step 3   Select feedback and assessment activities that support your learning goals.

Step 4   Select additional supportive learning activities for in class and homework.

Step 5   Integration progress check: Make sure your design has addressed your situational factors and that your assessment, feedback, and other learning activities truly support your learning goals. Adjust as needed before continuing.

Steps 6, 7, 8   Create a course structure, selecting or creating an instructional strategy, and then integrating the structure and strategy into an overall plan of learning activities.

Steps 9, 10, 11   Plan for your own evaluation of the course and teaching throughout the term, finalize what you will grade and how, review your materials with peers or mentors to debug possible problems.

And finally! Step 12   Finish by writing your plan for learning, including the basics of a syllabus, along with the instructions and rubrics for specific assignments.

You can use this chapter's Activity 6.1 to perform all these steps in more detail.

THE FOUNDATIONAL FIRST STEPS

In this chapter, we want to say more about the first five steps because they are critical to the task of creating an integrated, learning-centered course. One helpful tool for understanding the relationship between these key steps is Figure 6.2.

According to this model, the first step is to gather information about the "Situational Factors." Then use that information to make the three key decisions:

- What do you want your students to learn (what are your "Learning Goals")?
- What would they need to do to allow you (and the students) to know *whether they have achieved* the desired learning (what "Feedback and Assessment" procedures are needed)?
- How are they going to learn (what are the necessary "Teaching/ Learning Activities")?

FIGURE 6.2 The key components of Integrated Course Design.

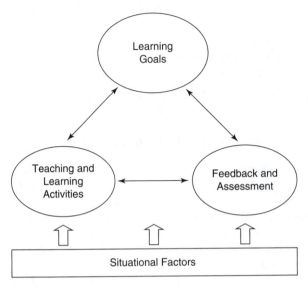

Then the three major components have to be aligned or "Integrated" (illustrated by the three bidirectional arrows), so that they support and reflect each other.

Step 1: What Situational Factors Should You Consider?

All instances of teaching take place within a context or situation: what is expected, who will be there, and what is the nature of the subject matter. Hence it is important for a teacher designing the learning to identify important aspects of the situation and gather a thorough amount of information about each of those aspects. The following five sets of potentially important factors offer the teacher a guide for which factors are important in a given situation:

1. *Specific context:*   This refers to such things as class size, the time structure, the level of the course, and how it will be delivered (live, online).
2. *Expectations of others:*   Frequently other individuals or groups have expectations for our courses that need to be considered. Some examples: a college or university's writing requirement, a profession's licensing exams.
3. *Nature of the subject matter:*   Some subject areas such as science, math, and engineering are more *convergent*; that is they strive to find the single, best answer. Others, such as the humanities, are more *divergent*, meaning that they strive to find multiple interpretations of given material.

4. *Nature of the teacher:*   How do we take advantage of our individual profile of skills, talents, and preferences? And improve these over time?

5. *Nature of the students:*   Our students have varied feelings about the subject matter, different levels of readiness, and different learning styles. The sooner these are understood and addressed, the better.

The specific context factors will always be important. The other factors are sometimes critical, sometimes not.

Step 2: How Do You Choose Learning Goals?

Decide what lasting impacts you most want for your students. Learning how to formulate significant learning goals for a course is key to getting beyond the limits of a content-centered course. For decades, teachers have used the Bloom (1956) taxonomy of educational objectives to assist in this task. Any taxonomy that continues to be used half a century after its publication is obviously a powerful idea. But in recent years, educators have been advocating important kinds of learning that do not fit into Bloom's taxonomy. It is for this reason that we offer a new "Taxonomy of Significant Learning" (Figure 6.3). Like Bloom's taxonomy, the taxonomy of significant learning identifies six major kinds of learning. Unlike Bloom's, these categories are not hierarchical but interactive. This means the more students learn any one of these goals, the easier it is to achieve the others.

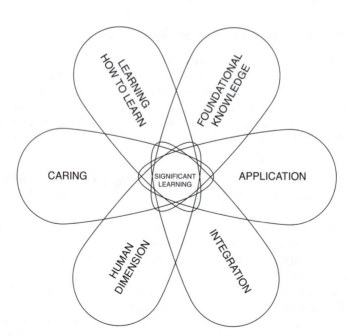

FIGURE 6.3   The interactive taxonomy of significant learning.

The general meaning of each kind of learning is as follows:

- *Foundational knowledge:*   What is it we want students to simply understand and remember, after the course is over? This is essentially the content of the course. Often the knowledge base of a field and subfield will have received considerable attention as it has, for example, in geography through the national standards and the *Geographic Information Science and Technology Body of Knowledge* (Geography Education Standards Project 1994; DiBiase et al. 2006).
- *Application:*   What do we want students to be able to *do*, by the end of the course?
- *Integration:*   What connections, relationships, or interactions should they be able to identify?
- *Human dimensions:*   What do we hope they will learn about *themselves*, or about interacting with *others*, as a result of their experiences in this course?
- *Caring:*   What new feelings, interests, or values would it be good for them to develop?
- *Learning how to learn:*   What can they learn now that will help them know how to continue learning, after the course is over?

Is it possible for a single course to strive for and achieve all six kinds of significant learning? Yes, some teachers have done it. But it requires making sure that you make the other components of the course strong enough to accomplish this: the assessment activities, the learning activities, and the teaching strategy.

Step 3: How Will You Know if Students Are Making Progress on the Learning Goals?

Measure what matters most. Once we know what we want students to learn, we need to determine how we would know whether, at the end of the course, they had succeeded in achieving these learning goals. In addition, throughout the course the students need to know how they are doing, and this is the function of feedback.

One of the challenges of doing good feedback and assessment is developing procedures that educate and go beyond simply auditing student learning, that is, did they "get it?" This broader form of assessment is called educative assessment. Doing it requires four actions by the teacher:

- *Use forward-looking assessment tasks:*   Rather than looking back at what was covered in a unit and asking "Did you get it?," create questions focused on situations and tasks that students might actually confront in the future that require them to use their new knowledge.

- *Develop clear criteria and standards:*   For every complex task that we ask students to do, we need to develop a clear vision for them of what constitutes good work on this task. Developing a set of focused, specific criteria for each assessment task and descriptions of high-quality and low-quality work (standards) for each criteria will greatly assist students in knowing what is expected of them.

- *Offer students the opportunity to engage in self-assessment:*   When students leave our courses and use their knowledge, they will need to know how to determine whether they are doing so in a competent way. If we give them multiple opportunities to self-assess their ability to apply their knowledge while in college, they will later be better able to evaluate the quality of their work and that of others.

- *Provide high-quality feedback:*   Whether we grade a student activity or not, students need feedback on the quality of their work. Our feedback will enhance learning much more if it is:
  - ○ *Frequent:*   One or two midterms and a final is not frequent; we need to assess learning more often throughout a course.
  - ○ *Immediate:*   If we wait too long, students only pay attention to the grade we gave them, not the feedback.
  - ○ *Discriminating:*   Our feedback needs to be based on clear criteria and standards.
  - ○ *Delivered in a user-friendly (or loving) manner:*   Offer constructive assessments—making students feel bad does not enhance their ability to learn from the feedback.

Step 4: How Can You Help Students Make More Progress on the Learning Goals?

Use the principles of holistic active learning to create or select activities that will enable students to achieve your learning goals. Challenging learning goals will definitely require us to follow the principles of active learning (for more on this topic, see Chapter 7 by Eric Fournier). Basically, this means the *set* of learning activities we use will need to include all three of the following kinds of learning activities (Figure 6.4):

- *Provide new information and ideas:*   Somehow students need to encounter new information and ideas about the subject. This can happen in multiple ways: readings, comments by the teacher, Internet web sites, and class discussions.

- *Call for students to do or observe something:*   A good set of learning activities has a strong experiential component. This may occur in real or authentic settings, for example, doing an authentic project or observing a real phenomenon in the field. Or it may happen in

FIGURE 6.4 A holistic view of active learning.

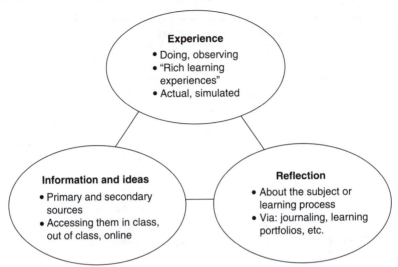

simulated, vicarious situations, for example, case studies, simulation exercises, role play, and movies.

- ***Prompt students to reflect on their learning:***   When students reflect on new information and ideas and on their experiences, they make learning meaningful to themselves. This can be done with one-minute papers, weekly journaling, or end-of-course learning portfolios.

When teachers look for ways to include all three of these forms of learning in each unit, the course becomes much more engaging, interesting, and capable of stimulating powerful kinds of learning.

Step 5: How Can You Make Sure Your Course Has an Integrated Design?

Verify that your choices support each other. The act of creating a more *differentiated* set of learning goals, assessment procedures, and learning activities is a major step toward creating a more powerful learning experience for students. But then we need to *integrate* these various components into a coherent whole. This is done in two ways, both of which are important.

First, we should create a three-column table for our learning goals, assessment procedures, and learning activities (Table 6.1). Start by filling out the left-hand column with each of the major goals for the course. Then, for each goal, fill out the rest of the row, that is, identify appropriate assessment procedures and learning activities for that learning goal.

This will ensure that we do more than give lip service to a set of noble learning goals; we will actually have the necessary assessment and learning activities built into the course that make them happen.

TABLE 6.1 Worksheet for integrating and aligning course components into a coherent whole.

Learning goals	Assessment procedures	Learning activities
1.		
2.		
3.		
4.		
5.		
6.		

Second, we need to create or select a powerful teaching strategy for the major units in the course. Our use of the term "teaching strategy" differentiates it from a "teaching technique," by which we mean a specific set of teaching/learning activities, for example, lecturing, whole-class discussions, and small group activities. A teaching strategy, for us, is a specific *combination* of teaching/learning activities in a particular *sequence*. A good teaching strategy is one that takes the right combination of activities and puts them in the right sequence. The "right" sequence positions each activity in the unit and course so that it builds on the activity that immediately preceded it and prepares students for what follows. It also concludes with an activity that ties the learning together and provides a culmination of the whole unit (or course).

There are some generic teaching strategies that teachers can use in a variety of courses, for example, team-based learning and problem-based learning. Or teachers can create their own strategy. When doing the latter, they need to make sure they use a good combination and a good sequence of learning activities.

WHAT ELSE SHOULD YOU CONSIDER BEFORE THE FIRST DAY OF CLASS?

Communicate your plan and expectations with others. When you design your courses as proposed here, you will need to think about how to alert other people to what you are doing. Generally speaking, it is a good idea to inform your department chair about what you are doing that is different from traditional teaching—and why (citing some of the examples presented here). You may also benefit from discussing this with any colleagues who are open-minded to what, in many colleges, is still considered innovative teaching instead of the norm.

Your students also need to know, from the first day of class, what the course learning goals are and how the course activities have been designed to help students move toward those goals. The learning activities you use in the first session, even as you provide an overview of the class, set student expectations for what and how you will present for the rest of the course. Students are far less resistant to nontraditional teaching if it begins in the first session. For example, put them in small groups or engage them in some reflective writing on the first day of class. Here are some things to think about doing that first day:

- Have students participate in at least one well-chosen active learning approach used within the first 20 min.
- Provide an overview of your learning goals, assessment activities, learning activities, and teaching strategy along with a rationale for why you made those decisions.
- Before the end of the first session, provide and review a clearly written syllabus setting out the schedule of required homework with due dates and grading rubrics; one possibility is to have a short quiz on the syllabus during your second session.
- Verify that their expectations are clear about what activities they will do prior to the next class session and what the consequences are for failure to complete those activities (e.g., "We will have a test on the readings without my lecturing on the material.").
- Give them a structured opportunity to connect with at least a few other students in the class.

Many teachers have found that, when they design their courses this way, students respond in a dramatically positive manner. They are more engaged in their learning, they work harder on their out-of-class activities, and they learn more and better. This does not mean that 100 percent of the students will achieve all the learning goals you have dreamed of. But the majority of your students in the majority of your courses will respond and learn much more than they do if you teach with just traditional methods.

Teaching is a very powerful profession to be in. The lifelong habits of learning you can give your students can affect the world well beyond your classroom. As Henry Adams said, you never know where your influence stops!

Additional Resources

Associations, conferences, and workshops

In recent years, various organizations have been hosting regional and national conferences focused on issues of college-level teaching and learning. Some prominent examples include The Lilly Conference at Miami University of Ohio, the Teaching Professor Conference, Regional Lilly Conferences, and the Annual Conference of the POD Network (the main professional association for faculty development).

The Association of American Geographers and National Council for Geographic Education conferences also offer opportunities for geographers to attend sessions specifically on the teaching of geography in higher education.

Fink, L. D. 2003. *Creating significant learning experiences: An integrated approach to designing college courses.* San Francisco: Jossey-Bass.

This book has more extensive comments on the basic model of Integrated Course Design described in this chapter.

Guides to good teaching, learning and assessment practices in geography under FDTL Guides. http://www.glos.ac.uk/gdn/publ.htm.

These guides are designed for geographers, but you can search online for subject-specific guides, teaching tips, and other materials.

Juwah, C., D. Macfarlane-Dick, B. Matthew, D. Nicol, D. Ross, and B. Smith. 2004. *Enhancing student learning through effective formative feedback.* http://www.heacademy.ac.uk/resources/detail/id353_effective_formative_feedback_juwah_etal.

This booklet provides seven principles of effective ways to structure feedback in a way that enhances learning, perhaps the best set of ideas on this topic.

Michaelsen, L. K., A. B. Knight, and L. D. Fink, eds. 2004. *Team-based learning: A transformative use of small groups in college teaching.* Sterling, VA: Stylus.

This book describes team-based learning, a powerful teaching strategy that "scales up," that is, it can be used with very large classes as well as smaller classes.

Professional Faculty Development at Your Institution

An increasing number of colleges and universities have faculty development workshops and other resources. Find out what is available at your school. At a minimum, you can usually find a colleague to talk with about your course and activity design decisions before you begin teaching.

Richlin, L. 2006. *Blueprint for learning: Constructing college courses to facilitate, assess and document learning.* Sterling, VA: Stylus.

This book contains a large variety of ideas that will help any course design process.

Teaching tips: a website at the University of Hawaii at Honolulu with an extensive collection of ideas and tools on teaching. http://honolulu.hawaii.edu/intranet/committees/FacDevCom/guidebk/teachtip/teachtip.htm.

Zubizarreta, J. 2004. *The learning portfolio: Reflective practice for improving student learning.* Bolton, MA: Anker.

This book describes the basic idea of a learning portfolio and then has multiple essays by people who have used them.

References

Barr, R., and J. Tagg. 1995. From teaching to learning: A new paradigm for undergraduate education. *Change Magazine,* November/December, 13–25.

Bloom, B. S., ed. 1956. *Taxonomy of educational objectives, handbook I: Cognitive domain.* New York: David McKay.

Bransford, J. D., A. L. Brown, and R. R. Cocking, eds. 1999. *How people learn: Brain, mind, experience, and school.* Washington, DC: National Academy Press.

Diamond, R. M. 1989. *Designing and improving courses and curricula in higher education: A systematic approach,* 1st ed. San Francisco: Jossey-Bass Publishers.

Diamond, R. M. 1998. *Designing and assessing courses and curricula: A practical guide,* rev. ed. San Francisco: Jossey-Bass Publishers.

DiBiase, D., M. DeMers, A. Johnson, K. Kemp, A. T. Luck, B. Plewe, and E. Wentz. 2006. *Geographic information science and technology body of knowledge,* 1st ed. Washington, DC: Association of American Geographers.

Fink, L. D. 2003. *Creating significant learning experiences: An integrated approach to designing college courses.* San Francisco: Jossey-Bass.

Geography Education Standards Project. 1994. *Geography for life: National geography standards 1994—What every young American should know and be able to do in geography.* Washington, DC: National Geographic Research and Exploration on behalf of the American Geographical Society, Association of American Geographers, National Council for Geographic Education, and National Geographic Society.

Saunders, P. 1980. The lasting effects of introductory economics courses. *Journal of Economic Education.* 12 (Winter): 1–14.

Svinicki, M. D. 2004. *Learning and motivation in the postsecondary classroom.* Bolton, MA: Anker.

Wiggins, G. and J. McTighe. 2005. *Understand by design,* 2nd. ed. Upper Saddle River, NJ: Pearson Education, Inc.

Zull, J. E. 2002. *The art of changing the brain: Enriching the practice of teaching by exploring the biology of learning.* Sterling, VA: Stylus.

CHAPTER 7

Active Learning

Eric J. Fournier

Active learning is not a set of techniques or a collection of methods; it is not limited by class size, student level or ability; and it is not bound to any particular subject. Rather, active learning is a philosophical approach to learning, an approach that shifts the educational focus from teacher to student. Active learning is fundamental to the vision of Integrated Course Design presented by Fink and Ganus in the previous chapter. For them, it is one of the four most important foundational steps in designing courses. Yet, the active learning Fink and Ganus advocate is often quite different than some of the models of teaching—such as lecturing—with which we as instructors are most familiar and most comfortable. As I discuss in this chapter, there are important theoretical and practical reasons for cultivating active learning in your classes and developing a repertoire of strategies suited to the learning outcomes of your classes.

At the theoretical level, active learning is based on the idea that learning is a process in which the learner actively constructs knowledge. Research in cognitive psychology suggests that methods involving students as active participants in the learning process produce better results. Gijselaers (1996) summarized the literature on active learning and educational theory and identified three key principles:

1. Learning is a constructive and not a receptive process.
2. Knowing about knowing (metacognition) affects learning.
3. Social and contextual factors influence learning.

The first principle means that learning occurs when new material can be linked to students' existing knowledge. The term *associative networks* is used to describe the mechanism by which new material is sorted and organized in the brain, but students often have poorly developed networks. For example, imagine you are a political scientist who just read a fascinating article about race-based redistricting in a southern state. You decide to share this with your American Government students, and they are underwhelmed. For a political scientist with a well-developed associative network, the article is linked to knowledge of party politics, redistricting, legislative processes, and political history. The article becomes fascinating because it is tied to existing knowledge. For students—especially introductory-level students—the article is not easily understood because they do not have well-developed prior knowledge of the subject.

Metacognition, the second key characteristic of active learning, refers to a student's ability to monitor the progress of his or her own learning through self-evaluation and, if needed, by adopting new techniques to learn material. This self-awareness is particularly useful when dealing with new information. Finally, the social and contextual factors of active learning can be introduced through realistic problems, group work, and application of material in different settings.

Traditional, lecture-based classes are safe. The pace, material, and content are managed and delivered by the teacher. For early career faculty, learning could be considered a high-risk educational strategy. Features of high-risk strategies include spontaneity, long time periods for resolution, and a lack of structure (Sutherland and Bonwell 1996). Philosophically, an instructor using various active learning approaches should fully understand the trade-offs and feel comfortable with covering a lesser amount of material. Many educators may agree that a course is only a starting point of learning, and much learning can take place after the class is over. After all, an instructor cannot tell students all they need to know about a subject in a single course. Introducing active learning can develop self-directed learning skills in students and enable them to become lifelong learners. From this point of view, the emphasis on learning skills and other learning issues is a worthy investment of class time and teachers' effort. As Thompson (1996) notes, many students—particularly in introductory-level classes—will never be practitioners in the field in question. But regardless of their career choice, they will need to be logical and scientific in their approach to challenges, evaluate the quality of a growing volume of data, and take positions on complex issues. This notion rings especially true in an age where most disciplines advance at a rapid pace and the shelf life of information is often quite brief (Cowdroy and Mauffette 1998).

WHY ACTIVE LEARNING?

The modern roots of active learning can be traced back to philosopher John Dewey who argued that education should be less about rote memorization and more about developing critical-thinking and problem-solving skills.

Dewey (1916) favored an approach that encouraged inquiry over passive learning:

> To realize what an experience, or empirical situation, means, we have to call to mind the sort of situation that presents itself outside of school; the sort of occupations that interest and engage activity in ordinary life. And careful inspection of methods which are permanently successful in formal education, whether in arithmetic or learning to read, or studying geography, or learning physics or a foreign language, will reveal that they depend for their efficiency upon the fact that they go back to the type of situation which causes reflection out of school in ordinary life. They give the pupils something to do, not something to learn; and the doing is of such a nature as to demand thinking, or the intentional noting of connections; learning naturally results.

Piaget's (1955) equally influential work dealt primarily with the development of learning in children, but a lot of what he discovered about how children learn can be applied to the college classroom. Piaget's theory is based on the idea that people build cognitive structures for understanding and responding to physical experiences within their environment. He found that a person's cognitive structure increased in sophistication with age, moving from a few innate reflexes to highly complex mental activities. In order for students to learn, teachers must emphasize the critical role that experiences—or interactions with the surrounding environment—play in student learning. Students must discover relationships and ideas in classroom situations that involve activities of interest to them, and that understanding is built up step by step through active involvement with the subject matter (Piaget 1955).

Finkel (2000, 8) puts Piaget's theory into practice in his book *Teaching with Your Mouth Shut* where he argues that "good teaching is the creating of circumstances that lead to significant learning in others." He rejects the idea of teaching as telling and asks instructors to shift the focus of their classes from themselves to their students. This idea can be traced to Freire who argued against the "banking model" of education (in which a student's head is treated as an empty vessel waiting to be filled with "deposits" of wisdom from a teacher) in favor of a democratized model of education where students and teacher were partners in the educational process (Freire 2000; Merrett 2000; Griffiths 2005). Finkel (2000, 53) echoes Dewey when he writes: "The disequilibrium that results from an interruption in our ongoing interaction with the world is what motivates learning." In an active learning setting, students are confronted with a problem or challenge. They have a need to solve that problem and thus become interested in the issues surrounding that problem.

One of the most compelling arguments in favor of active learning is how well it helps students reach the outcomes we set for our classes. For example, if we examine Bloom's (1956) influential taxonomy of educational outcomes of the cognitive domain, it is easy to see that active learning can support all six outcomes: knowledge, comprehension, application, analysis, synthesis, and evaluation (Krumme 2005). The latter outcomes—those often viewed as "higher-order" thinking skills—are especially suited to active learning strategies. Less attention has focused on the subsequent work of Bloom and his colleagues to characterize the educational objectives in the affective domain (Krathwohl, Bloom, and Masia 1964). Yet, *affect*—feeling, emotion, or engagement with a question, problem, or issue—is often crucial to long-term learning. Again, active learning is a means to spur the sort of interest, emotion, and engagement that supports long-term learning.

But our courses aim for outcomes that go beyond the cognitive and even affective domains of Bloom's taxonomies. Angelo and Cross's (1993, 13–23) Teaching Goals Inventory lists, for instance, over fifty possible outcomes grouped into six broad categories: (1) higher-order thinking skills; (2) basic academic success skills; (3) discipline-specific knowledge and skills; (4) liberal arts and academic values; (5) work and career preparation; and (6) personal development. Fink's (2003) taxonomy of significant learning presented in the previous chapter groups outcomes somewhat differently into: (1) foundational knowledge; (2) application; (3) integration; (4) caring; (5) human dimension; and (6) learning how to learn. My point is that no matter how we categorize our goals, active learning can be a key to reaching them.

A final theoretical justification for active learning comes from Kolb (1984, 1985) and his research into experiential learning theory and learning styles. Kolb argued that not one, but four types of experience need to be interlinked for learning to occur: (1) concrete experience; (2) reflective observation; (3) active experimentation; and (4) abstract conceptualization. Students often have preferences (or learn more readily) from one of these four types of experiences—preferences Kolb termed "learning styles"—but the fact remains that all four work together to produce significant learning. He urges teachers, when designing courses, to incorporate methods that take these various styles into account. Ideally, these would include some element of active learning (Healey and Jenkins 2000).

Active learning takes many forms and each of its variants comes with an array of supporting literature and a cohort of zealots eager to promote the specific method as the pinnacle of pedagogy. For faculty new to active learning, the volume of material can be intimidating, but essentially it is about learning by doing (Healey and Roberts 2004). Sociologists, for example, should examine the various active learning methods and ask "How will this help my students learn sociology?" Choose methods that help meet the learning goals established for the class. These might involve class discussion, role playing, problem-based learning, field work, or a computer simulation.

Active learning can also supplement a lecture-based class. Donald Bligh (2000) suggests several ways of injecting an active approach into lectures. These include buzz groups, where students gather in small groups to discuss an issue introduced in lecture or to accomplish some sort of tasks related to the class topic; and horseshoe groups, where two or more buzz groups gather together to synthesize small group issues. These group efforts may be followed by controlled discussion in which a spokesperson reports out to the class (Table 7.1).

TABLE 7.1   A sample of active learning strategies.

Think/pair/share

Have students turn to someone near them to summarize what they are learning, to answer a question posed during the discussion, or to consider how and why and when they might apply a concept to their own situations. The procedure is as follows: (1) individuals reflect on (and perhaps jot notes) for one minute in response to a question; (2) participants pair up with someone sitting near them and share responses/thoughts verbally for two minutes; (3) the discussion leader randomly chooses a few pairs to give thirty second summaries of ideas.

One minute paper/free write

Ask participants to write for two to three minutes on a topic or in response to a question that you have developed for the session. The moments of writing provide a transition for participants by bringing together prior learning, relevant experience, and new insights as a means of moving to a new aspect of the topic.

Scenarios/case studies

Provide students with an example of a topic being covered in the discussion or lecture. Participants discuss and analyze the scenario/case, applying the information covered in a presentation to some situation they may encounter outside of the class. Students can briefly present their findings to other small groups or to the whole group.

Corners

The instructor places a flipchart or sheet of poster paper with a different question in each corner of the room. Groups of students move from corner to corner and discuss answer(s) to each posed question. The groups develop a consensus and write their answer directly on each flipchart. When the flipchart has an answer already written by a previous group, the next group revises/expands/illustrates that response with additional information, if possible.

Note check

Students pair with a partner/small group to briefly (two to five minutes) share notes. They can clarify key points covered, generate and/or resolve questions, generate a problem to solve, solve a problem posed by the instructor, or write a paragraph synthesizing key ideas as set out in partner's notes.

TABLE 7.1 Continued

Jigsaw teamwork

A jigsaw is an active learning exercise in which a general topic is divided into smaller, interrelated pieces; each member of a team is assigned to read and become an expert on a different piece of the puzzle; then, after each person has become an expert on his/her piece of the puzzle, he/she teaches the other team members about that puzzle piece; and, finally, after each person has finished teaching, the puzzle has been reassembled and everyone in the team knows something important about every piece of the puzzle.

Source: Adapted from University of Minnesota, Center for Teaching and Learning. 2006. Scenes from a classroom: Making active learning work. http://www1.umn.edu/ohr/teachlearn/tutorials/active/strategies.html (last accessed 6 February 2008).

DO NOT REINVENT THE WHEEL

A colleague of mine once said that the secret to teaching success was to lie, cheat, and steal. I will leave the lying and cheating for another time, but by encouraging faculty to steal, he meant taking the good ideas that others have developed and using them for your own classes. There is no need for most instructors to develop active learning materials from scratch because there are a lot of high-quality activities available which can be easily adapted for a variety of classroom settings. If you prefer, instead of calling it stealing, we can call it the diffusion of good educational practice!

See "Additional Resources" below for a sample of some of the available resources that can provide you with more advice about active learning. In addition to the general resources, I have included some from geography to provide an overview of good discipline-specific resources. In addition to these web resources, there is a range of books and printed materials that may prove useful (McGlynn 2001; Weimer 2002; Barkley, Cross, and Major 2004; Healey and Roberts 2004; Nilson 2007).

You will notice that some of the links in the list are for "problem-based" learning or "inquiry" learning (Spronken-Smith 2005). Active learning is today really an umbrella term for a number of techniques all rooted in the same constructivist theory. Although the particular techniques vary somewhat in design, approaches of discovery learning, experiential learning, problem-based learning, and inquiry-based instruction all fit under the active learning umbrella.

You may not have to reinvent the wheel to begin experimenting with active pedagogy, but it is important to guide the process in the right direction as you get started. As I argued earlier, active techniques can be adapted and customized to support many different learning outcomes— but it is up to you to make sure that your activity and outcome are aligned and integrated. So avoid using ideas and examples without carefully considering how they fit into your overall course plan. In the next section, for example, I show how the nature of the learning activity can be subtly shaped to focus on particular learning outcomes.

EXAMPLES OF ACTIVE LEARNING AND HOW THEY CAN BE REFINED

In order to illustrate the use of active learning, I have provided examples of active learning for several different classes. Before you read further in this section, think back to Bloom's taxonomy and try to classify each of the activities (Tables 7.2–7.4) according to the six levels of cognitive processes at which they are aimed (knowledge, comprehension, application, analysis, synthesis, and evaluation). Later, please visit the book's web site and complete Activities 7.1 (a brainstorming exercise for developing active learning strategies) and 7.2 (a template-based approach to designing your own active learning exercises).

TABLE 7.2 **The first of three examples of a classroom activity focusing on the concept of albedo. Each activity focuses on engaging learners at different cognitive levels.**

Albedo and net radiation

Albedo: the percentage of shortwave radiation scattered upward by a surface.

The albedo is calculated by dividing incoming shortwave radiation from the sun by outgoing (or reflected) shortwave radiation from the surface:

$$\text{Albedo} = K{\uparrow}/K{\downarrow} * 100$$

$K{\downarrow}$ = Incoming shortwave radiation from the sun
$K{\uparrow}$ = Reflected shortwave radiation from the surface

Different surfaces have different reflective properties:

Forest	5–10%
Grass	20–25%
Dry sand	20–30%
Old snow	50–60%
Fresh snow	80–90%
Water	3–5%
Water (sunrise and sunset)	50–80%

1. What type of surface is most reflective? Name another highly reflective surface (not from the list).
2. What type of surface has the lowest albedo? Name another surface with a low albedo (not from the list).

TABLE 7.2 Continued

3. Near the poles there is always a low percentage of solar absorption. What kinds of surfaces account for this?
4. Briefly describe the seasonal variation in albedo to be expected in a broadleaf deciduous forest in New England (no leaves in winter, snow cover).

Adapted from Suckling and Doyon (1991).

The concept of albedo is a typical topic in most intro-level physical science classes. Many classes might begin with a definition—*Albedo is the percentage of insolation reflected by a surface.* Many teachers supplement that definition with examples (*light versus dark surfaces, snow versus forest*) or illustrations. Others might employ anecdotes—*a lighter colored roof might save on cooling bills in summer.* Some others might introduce the latest research on the subject (*melting polar ice will decrease planetary albedo and lead to an accelerated cycle of global warming*). An active learning approach to the concept of albedo invites students to investigate the concept on their own.

In Table 7.2, students complete a worksheet that gives them an opportunity to think about the concept of albedo. The first two questions simply reinforce the topic, but the last two questions ask students to apply the concept to a real-world scenario. One of these questions might be delivered in the form of a minute paper or a think/pair/share and can be easily used in large classes.

In this next example (Table 7.3), students are asked to apply the concept of albedo in an exercise involving analysis and synthesis of diverse types of information. Students are asked to calculate an estimated albedo for their campus. Materials for this might include a campus map, air photo or satellite image (perhaps in different seasons), an overlay grid for calculating surface area, colored pencils for classification, and perhaps a basic land-use classification software program.

The final example (Table 7.4) asks students to gather their own data to prove or disprove the hypothesis that dark surfaces absorb more radiation than lighter surfaces. This aims for Bloom's evaluation objective in that students are asked to make judgments about the value of ideas and evidence and to select the most effective experiment. Teams of students are provided with handheld, infrared thermometers to record surface temperatures, and they are given a template to record their work. Students design their own experiment using the thermometers to test the hypothesis. Once they devise a strategy, they come up with a methodology and then go outside to gather data. Once they have gathered data they return

TABLE 7.3   The second of three examples of a classroom activity focusing on the concept of albedo.

Lab exercise: Campus albedo

Introduction: The diagram in your text shows the albedos for various surfaces. Your task is to examine the albedos for different surfaces and then:

A. Use the map and your powers of observation to identify the various surfaces on campus
B. Estimate the percentage of the campus covered by each surface
C. Match each surface with the average albedo shown in Table 7.2
D. Calculate the average albedo for the campus

Type of surface	Albedo	% of surface on campus	

1. How did you identify the various surfaces on campus?
2. How did you estimate the percentage covered by each surface?
3. How did you calculate average albedo for the campus?

to the classroom to organize and analyze their results and then either accept or reject the hypothesis. Students who have done this activity used many creative data sources—parking lot versus grass, black cars versus white cars, and one group got people with different colored shirts to lie still in the sun for ten minutes absorbing (or reflecting) various amounts of insolation.

TWO EXAMPLES OF PROBLEM-BASED LEARNING

The hybrid model of problem-based learning (PBL) is a popular method in many disciplines. While active learning can be used in any discipline, geography, with its focus on complex, interdisciplinary, multiscale challenges, is particularly well suited to this approach. However, the method does introduce some issues that can affect its effectiveness. Some of the concerns of using this method in an intro-level course include: lack of student background in the subject and experience doing collaborative and cooperative

TABLE 7.4 The third of three examples of a classroom activity focusing on the concept of albedo.

Physical geography: Albedo lab

Hypothesis: Dark surfaces absorb more incoming solar radiation than lighter surfaces

We will prove/disprove this by:

Methods:

Data:

work, time constraints, and a lack of student research experience. Activity 7.3 is a guide to creating effective PBL activities in your courses.

In this example of PBL in geography, students assume the role of a poor, Mexican farmer living in rural, northern Mexico. Each group member chooses a different option to research: *Migrate to Mexico City, Migrate to Ciudad Juarez, Migrate to the U.S., Migrate to Monterrey, or Stay in rural Mexico.* Students research their option as individuals—noting both positive and negative aspects of the option—then reconvene as a group to vote on the best option. Later the groups gather together in a class discussion to evaluate positive and negative aspects of each option and to decide on the best option. By completing this exercise, students learn about migration, conditions in rural Mexico, the physical geography of Mexico, and develop a basic under-standing of economic issues such as North American Free Trade Agreement.

In order to address the student's lack of background in the subject, the unit begins with a short lecture on the basics of Mexican geography, intro-ducing the country, and providing a common background for all students. The introductory lecture also examines student stereotypes concerning the country. For example, many students view Mexico as a small, impoverished country. A map of Mexico is overlaid on a map of the U.S. and shows Mexico stretching from Oregon to Florida. Comparative economic data are then pre-sented that show that, by world standards, Mexico is a solidly middle-class country in terms of per capita income and manufacturing output. The chap-ter on Mexico from the text is also assigned. These steps help assure that all students have been exposed to the same background materials as they begin their assignments. In this case, the module involves individual work, but the assignment cannot be completed without group cooperation. In addition, clearly defined products help students work together. In this case, a short paper is required from each individual and the final product from each group is a migration decision supported by facts. In keeping with the open-structured nature of PBL, there are no right or wrong answers. Each migration decision is valid as long as it is supported by valid information and exhibits sound logic (Fournier 2003).

In another example of PBL that I have employed, students assume the role of a junior executive for a multinational conglomerate. They monitor world affairs in a particular region of the globe and prepare a weekly report for their boss (played by the teacher). These regional expert reports extend throughout the semester and students work and are assessed as individuals rather than as groups. Understanding current events is an important part of geographic literacy, and each student in the class is assigned a different world region to monitor. Executive summaries of the region's events are due each week during the semester based on this scenario:

> You may imagine that your report is intended for the CEO of an international corporation. The CEO is too busy to constantly mon-itor world affairs, so she has a staff to do the work. Each week you should prepare an executive summary of no more than one page. Of course, it must be typewritten. An outline style is preferable to a more narrative style. Remember this is just a summary, not an in-depth analysis. Some weeks there may not be much news from your region. In that case, a one-page report on some aspect of your region's economy, culture, politics, environment, or (almost) anything else is acceptable. Sources of information should be care-fully noted. As the semester progresses regional experts may be asked to share some of their expertise with their classmates.

This exercise develops many essential skills. Each week students locate appropriate sources for their respective regions, monitor those sources, prepare a written report summarizing events for their region, demonstrate a familiarity with their region's human and physical geography, and eventually develop an appreciation for the complexity of events in their region. In addition, successful students cite proper sources for their regional expert reports, summarize a week's events in a single page, evaluate and rank a week's events in order of significance, collect a number of reports from a variety of sources, and perform an editorial function by choosing five of the most significant events of the past week for their reports. Finally, students evaluate the quality of the information, especially when presented with conflicting facts or interpretations (Chapman, Keller, and Fournier 2002).

CONCLUSION

There are both theoretical and practical justifications for using active learning in college classes, and any instructor who is serious about his or her students learning the subject should begin to experiment with active learning strategies such as those discussed in this chapter. Effective use of active learning strategies does not require the total overhaul of a class or the complete abandonment of traditional methods. Even a sprinkling of active learning into an existing class can help raise student interest and attention. To aim for significant learning it is always important to make sure that your choice of activities is always carefully aligned with the course goals and objectives. As Fink and Ganus stress in Chapter 6, holistic active learning always provides learners with new information and ideas, an opportunity to do or observe something, and time to reflect on their learning. The online activities associated with this chapter can help you put these principles of active learning into practice.

Additional Resources

Many of these resources provide examples and ideas which can be readily adapted to a wide range of courses.

Interdisciplinary Resources

Active Learning with PowerPoint, http://www1.umn.edu/ohr/teachlearn/tutorials/powerpoint/

Active Learning for the College Classroom, http://www.calstatela.edu/dept/chem/chem2/Active/

Active Learning Online, http://www.acu.edu/cte/activelearning/

Samford University, Center for Teaching, Learning, and Scholarship, Problem-Based Learning, http://www.samford.edu/ctls/problem_based_learning.html

The Active Learning Site, http://www.active-learning-site.com/

Thirteen Online Education, Inquiry-based Learning, http://www.thirteen.org/edonline/concept2class/month6/index.html

University of Gloucestershire, Center for Active Learning, http://www.glos.ac.uk/ceal/
University of Illinois, Urbana-Champaign, Inquiry Page, http://www.inquiry.uiuc.edu/inquiry/definition.php
USGS Science Resources for Undergraduate Education, http://education.usgs.gov/common/undergraduate.htm

Discipline-specific Resources, Examples for Geography

Center for Global Geography Education, http://www.aag.org/education/center.
Earth Observatory, http://earthobservatory.nasa.gov
Geography Discipline Network, http://www.glos.ac.uk/gdn
Global Change in Local Places, http://www.aag.org/gclp/gclpnew.html
Hands-On! Human Dimensions of Global Change, http://www.aag.org/HDGC/Hands_On.html
Mission Geography, http://geog.tamu.edu/sarah/mg.htm
Talessi Project: Teaching and Learning at the Environment-Science-Society Interface, http://www.greenwich.ac.uk/~bj61/talessi/
Strategies and Ideas for Active Learning, http://www2.una.edu/geography/Active/strategi.htm

References

Angelo, T. A., and K. P. Cross. 1993. *Classroom assessment techniques: A handbook for college teachers*. San Francisco: Jossey-Bass.
Barkley, E., K. P. Cross, and C. H. Major. 2004. *Collaborative learning techniques*. San Francisco: Jossey-Bass.
Bligh, D. A. 2000. *What's the use of lectures?* San Francisco: Jossey-Bass.
Bloom, B. S. 1956. *Taxonomy of educational objectives, handbook I: The cognitive domain*. New York: David McKay.
Chapman, D. W., G. E. P. Keller III, and E. J. Fournier. 2002. *Implementing problem-based learning in the arts and sciences*. Birmingham, AL: Samford University Press.
Cowdroy, R., and Y. Mauffette. 1998. Thinking science? Or science thinking? The challenge for science education. In *Themes and variations in PBL*, Vol. 1, eds. J. Conway and A. Williams, 30–39. Callaghan, NSW, Australia: Australian Problem Based Learning Network.
Dewey, J. 1916. *Democracy and education*. New York: Macmillan. http://www.ilt.columbia.edu/publications/dewey.html (last accessed 6 February 2008).
Fink, L. D. 2003. *Creating significant learning experiences*. San Francisco: Jossey-Bass.
Finkel, D. L. 2000. *Teaching with your mouth shut*. Portsmouth, NH: Heinmann.
Fournier, E. J. 2003. World regional geography and PBL: Using collaborative learning groups in an introductory-level world geography course. *Journal of General Education* 41 (4):293–305.
Freire, P. 2000. *Pedagogy of the oppressed*, 30th anniversary ed. New York: Continuum International Publishing Group.
Gijselaers, W. J. 1996. Connecting problem-based practices with educational theory. In *Bringing Problem-Based Learning to Higher Education: Theory and Practice*, eds. L. Wilkerson and W. J. Gijselaers, 13–21. San Francisco: Jossey-Bass.

Griffiths, H. 2005. Funky geography: Paolo Freire, critical pedagogy, and school geography. M.Sc. thesis, University of Birmingham, School of Geography, Earth and Environmental Sciences. http://www.gees.bham.ac.uk/downloads/gesdraftpapers/helengriffiths-thesis.pdf (last accessed 6 February 2008).

Healey, M., and A. Jenkins. 2000. Kolb's experiential learning theory and its application in geography in higher education. *Journal of Geography* 99 (5):185–95.

Healey, M., and J. Roberts. 2004. *Engaging students in active learning: Case studies in geography, environment, and related disciplines*. Cheltenham, U.K.: University of Gloustershire, Geography Discipline Network.

Kolb, D. A. 1984. *Experiential learning: Experience as the source of learning and development*. Englewood Cliffs, NJ: Prentice-Hall.

———. 1985. *Learning style inventory*, rev. ed. Boston: McBer.

Krathwohl, D., B. S. Bloom, and B. B. Masia, eds. 1964. *Taxonomy of educational objectives, handbook II: Affective domain*. New York: David McKay.

Krumme, G. 2005. *Major categories in the taxonomy of educational objectives (Bloom 1956)*. http://faculty.washington.edu/krumme/guides/bloom1.html (last accessed 6 February 2008).

McGlynn, A. P. 2001. *Successful beginnings for college teaching*. Madison, WI: Atwood Publishing.

Merrett, C. 2000. Teaching social justice: Reviving geography's neglected tradition. *Journal of Geography* 99 (5):207–18.

Nilson, L. B. 2007. *Teaching at its best: A research-based resource for college instructors*. San Francisco: Jossey-Bass.

Piaget, J. 1955. *The child's construction of reality*. London: Routledge and Kegan Paul.

Spronken-Smith, R. 2005. Implementing a problem-based learning approach for teaching research methods in geography. *Journal of Geography in Higher Education* 29 (2):203–21.

Suckling, P., and R. Doyon. 1991. *Studies in weather and climate*, 3rd ed. Raleigh, NC: Contemporary Publishing Company.

Sutherland, T. E., and C. C. Bonwell, eds. 1996. *Using active learning in college classes: A range of options for faculty*. San Francisco: Jossey-Bass.

Thompson, A. M. 1996. Problem-based learning in a large introductory geology class. *About teaching: A newsletter of the University of Delaware Center for Teaching Effectiveness*. Newark: University of Delaware. http://www.udel.edu/pbl/cte/spr96-geol.html (last accessed 6 February 2008).

University of Minnesota, Center for Teaching and Learning. 2006. Scenes from a classroom: Making active learning work. http://www1.umn.edu/ohr/teachlearn/tutorials/active/strategies.html (last accessed 6 February 2008).

Weimer, M. E. 2002. *Learner-center teaching: Five key changes to practice*. San Francisco: Jossey-Bass.

CHAPTER 8

Advising Students

Fred M. Shelley and Adrienne M. Proffer

It is a warm, humid April afternoon near the end of the spring semester. You are just back from a national meeting, and there is a lot to do before finals week, which is quickly approaching. Returning to your office from lunch, you find an e-mail from Megan, who is a geography major finishing her junior year at your university. Megan wants to go to graduate school after completing her undergraduate degree next year, and she wants your guidance about the application process. As you reply, you hear a knock on your door. You look up to see José, who is a student in one of your undergraduate classes. José has come to tell you that his father was laid off from his job, and that he has decided to drop out of college in order to go to work to help support his family. A few minutes later, the phone rings. On the line is Kelsey, your reliable and normally cheerful graduate assistant. Between sobs, Kelsey asks if she can speak with you as soon as possible about an urgent personal issue.

Megan, José, and Kelsey are among the many students who seek and rely upon advice from faculty members everyday. For these and other students, "good advising may be the single most underestimated characteristic of a successful college experience" (Light 2001, 80). How can you help these students and, at the same time, avoid increasing the already considerable demands on your time? Faculty members in geography and other disciplines must deal with many pressures in their efforts to succeed professionally in publication, teaching, and service. These expectations often mean that one of the most important aspects of a faculty member's position—the advising of graduate and undergraduate students—is often overlooked or ignored (Hennessy 2004). Yet, the advice that you can give Megan, José, Kelsey, and other students may make a tremendous difference in their professional and personal lives.

How can you advise students effectively, responding to their needs, while still maintaining a balance between helping them and achieving other professional goals? In this chapter, we discuss the advising process, present some principles associated with the effective practice of advising, link these principles to the literature on academic advising, and present in Activity 8.1 some advising-related scenarios for you to analyze using these principles.

We concur with Chickering's (1994, 50) view that:

> The fundamental purpose of academic advising is to help students become effective agents for their own lifelong learning and personal development. Our relationships with students—the questions we raise, the perspectives we share, the resources we suggest, the short-term decisions and long-range plans we help them think through—all should aim to increase their capacity to take charge of their own existence.

This is what is termed "developmental advising" insofar as it "is concerned with elevating the most routine advising practices, such as answering students' everyday questions to an opportunity for student learning and personal development" and uses "interactive teaching, counseling, and administrative strategies to assist students to achieve specific learning, developmental, career, and life goals. These goals are set by students in partnership with advisors and are used to guide all interactions between advisor and student" (Creamer and Creamer 1994, 17, 19).

Advising is, of course, related to mentoring and coaching; all three activities (and definitions) overlap to some degree. Mentoring, as a form of advising, tends to focus on professional situations in which an experienced, senior person helps a less-experienced person (the mentee or protégé) to advance in a particular career or job by offering counseling, training, discussion, and other assistance (Moss et al. 1999; Hardwick 2005; Hardwick and Shelley 2007). In contemporary usage, coaching has come to be used in some contexts as a synonym for advising—but not in the sense of an expert coach standing on the sidelines, always ready with needed answers and advice. Rather, coaching means helping others reach their goals not by offering advice but by giving encouragement and by helping the advisee clarify goals and solve problems. Although advising, mentoring, and coaching are related, we focus on academic advising in this chapter.

THE IMPORTANCE OF THE ADVISING PROCESS

The psychologist Drew Appleby recognized that effective advising is often linked to effective teaching. "When I began my academic career, I was certain that teaching and advising were different activities. I believed that teaching consisted of giving lectures to groups of students in the classroom . . . I believed that advising consisted of meeting individually with my psychology major advisees in my office to help them choose classes . . ." (Appleby 2001). With experience, Appleby recognized that advising and teaching are inseparable. Both require effective preparation, positive attitudes, and genuine concern for students. Both are predicated on the underlying objective of helping students succeed not only in mastering intellectual material but also in assimilating this material in planning for their lives and careers.

In today's fast-paced academic world, we often overlook advising. Why is this so? For starters, many faculty members face relentless pressure to publish and to obtain external funds to support research activities. Ph.D. students who seek tenure-track positions must also demonstrate their ability to succeed in publication. Tenure-track faculty members who spend a lot of time with students are sometimes criticized as insufficiently engaged in research and the search for external funds. Part-time adjunct professors, lecturers, and instructors face many of the same time pressures, especially given that many are paid on a per-course basis by the number of courses they teach, with no additional compensation or credit for advising.

Over the years, these pressures along with changes in institutional structure have affected the advising process, frequently separating students from professors. Advising is often limited to ensuring student compliance with often arbitrary standards and requirements imposed by university administrations. These advising efforts are often delegated to professional advisors, who may or may not have backgrounds in the student's major. These professional advisors are charged with ensuring student compliance with departmental, college, and university degree requirements. However, many professional advisors have little or no knowledge of disciplinary cultures and many have little time to deal with the needs of individual students. The work of professional advisors can complement, but cannot replace the interaction between professors and students that is at the core of academic life.

Academic institutions are, however, beginning to recognize the important role of the advising process as part of faculty performance evaluations. For example, Clemson University has included effective advising as a "significant criterion" for tenure and promotion: "Guidelines shall be developed for the inclusion of effective academic advising as a significant criterion for tenure, promotion, and other personnel actions for all faculty and staff whose duties include advising or the supervision of advising" (Clemson University 2001). Pennsylvania State University's promotion standards also refer to the importance of advising:

> Promotion and tenure decisions shall be based on these three criteria . . . The Scholarship of Teaching and Learning—ability to convey subject matter to students; demonstrated competence in teaching and capacity for growth and improvement; ability to maintain academic standards, and to stimulate the interests of students in the field; **effectiveness of counseling, advising and service to students**. . . (Penn State 2003, emphasis added).

Not only is effective advising important in and of itself, but its practice is becoming identified as a key component in tenure and promotion decisions.

EFFECTIVE ADVISING PRACTICE

Where should Megan go to graduate school? To what extent should you discourage José from giving up his dream to complete a college education? Why was Kelsey crying when she called you? How can you help her resolve her issues? Approaches to answering these and other questions involving interaction between faculty and students are associated with effective academic advising practice.

In advising students, it is important for advisors to recognize that many undergraduate and graduate students are young adults who are going through numerous transitions in their personal and professional lives. Persons in their late teens and twenties are making the sometimes uneasy transition from childhood to adulthood. They are making important decisions about their futures, establishing new relationships, marrying, and having children. Given the transitions associated with young adulthood, the fact that students frequently change majors or thesis topics should come as no surprise.

Advising is intimately related to life planning and career development and can be done in both formal and informal settings. As Gregory (2002) says, "if we pause once in a while to remember with whom we deal—human beings at their own crossroads, both academic and personal—it's not difficult to see that academic advising is teaching, personal counseling, and career counseling." In that light, effective advisors understand that today's world is one of rapid change. How will terrorism, global warming, environmental degradation, energy shortages, and other social and environmental issues impact the world in which today's students will live and work? Studies show that the average American will change careers three or four times over the course of his or her lifetime. Moreover, the vulnerability of many students is enhanced by other potentially negative influences—peer pressure, mass media, drug and alcohol abuse, and so on. Thus, not only are college students preparing for careers at a time when changes are frequent, but the pace of change throughout society continues to accelerate (Friedman 2005).

Much has been written about effective academic advising, and we summarize some important resources below in the section "Additional Resources." Perusal of the advising literature, along with experience, suggests a few important guidelines. Next, we identify and briefly discuss seven principles. We then consider application of these principles to the unique position of geography as a discipline.

Know the Purpose of Advising

The National Academic Advising Association (NACADA) provides a set of "core values" when it comes to advising. The first value states that "advisors are responsible to the individuals they advise." NACADA explains that "advising, as part of the educational process, involves helping students develop a realistic self-perception and successfully transition to the postsecondary institution. Advisors encourage, respect, and assist students in establishing their goals and objectives" (NACADA 2004). In this way, effective advising links quality academic performance to life planning. In the short run, the job of the advisor is to facilitate successful completion of a degree program. In the long run, the successful completion of one's degree requirements is associated intimately with planning for life itself. Completion of the degree represents a beginning, not an end. It is no coincidence that the completion of degree requirements are recognized in "commencement" ceremonies—that is, they celebrate the beginning of postgraduate life, rather than the end of an education.

Effective advising involves two-way communication. Advisors are not dictators. The advisor suggests courses of action that he or she feels will be beneficial to helping the student achieve particular goals. In many cases, however, good advisors refrain from making specific suggestions. It is often a good idea to play a more neutral role, allowing the student to talk out his or her problem. For example, suppose students are undecided between graduate school and going into the workforce after graduation. Rather than steer students in one direction or the other, ask them to articulate what they feel are the advantages and disadvantages associated with each alternative. As they spell out these advantages and disadvantages, they are more likely to see for themselves which alternative is more appropriate given their particular circumstances.

Good advice is given with what you feel is the student's best interests at heart, and it is backed up by sound reasoning, logical thinking, and consideration of consequences associated with proposed courses of action. Yet a student may have good reasons to decline your advice. For example, you may feel that a student might excel in a particular career, but the student decides to move in a different direction. Do not take rejection of your advice personally.

Know the System

Many if not most undergraduate and graduate programs have procedures by which formal advising functions are assigned or delegated to individual faculty members. At the graduate level, the chair of a student's thesis or dissertation committee is generally referred to as an "advisor" or "research advisor" and is charged with directing the student toward completion of the research project. Other faculty members serve as members of the student's thesis or dissertation committee, playing a less-active, yet still significant, role in the undertaking and completion of the research.

The direction of a thesis or dissertation project is a very important, but by no means sole aspect of effective advising. Effective advisors not only provide guidance about the research projects themselves, but they also provide guidance to their students about life planning in general. Likewise, effective advising cannot be limited to the four walls of academic institutions. Field trips, social events, meetings and conferences, and other settings also provide opportunities for advising. Every successful teacher is aware of and takes advantage of "teaching moments" in both formal and informal settings.

Effective advisors have a thorough understanding of the departmental and university policies regarding degree programs. This does not mean that the advisor has to memorize every departmental or university policy or procedure, but the advisor should be prepared to refer questions about policies and procedures to those who know or can discover the answers. It is also important not to promise more than you know you can deliver. For example, you may feel that a university-wide general education requirement may be inappropriate for a particular advisee and you may advise the student to request a waiver of this requirement. Unless you have the authority to waive the requirement yourself, don't promise the student that it will be waived. Rather, explain to the student that you will do your best to advocate waiving the requirement, but that the decision as to whether or not the requirement is waived is beyond your control.

Effective advisors also participate in student-oriented activities at the departmental and university levels. Doing so helps the advisor get to know more fully the "system" of which students are a part. For example, De Sousa (2005) suggests that advisors should "fully participate in orientation and first-year programs." Going to freshman orientation, for example, will remind the advisor of what it was like to be a first-year college student. It will help you to get to know the students and provide an opportunity for the students to recognize you as an interested and approachable member of the department. Similarly, volunteering to participate in departmental or university orientation programs for graduate students provides insight into the issues that these more mature students are facing.

Listen Effectively

A student's problem may seem trivial to you, but it is important enough for the student to seek your advice about it. Thus, it is important to listen closely and effectively to the student.

Hennessy (2004) advocates the application of the FISH! principles to academic advising. The FISH! principles include choosing the right attitude, playing, being there, and making the customer's day. This philosophy came about as a result of the former working conditions at the Pike Place Fish Market located in Seattle, Washington. Implementing these principles turned a seemingly dreary job into the now world-famous place where the workers "playfully" throw fish back and forth, engaging the customers and passersby. Those who have studied these principles believe that they can be applied in any work setting, and the use of these principles in an advising setting is no exception (Lundin, Paul, and Christensen 2000). Hennessy states that "We make a personal choice every day whether we are going to be passionate and energetic or negative and condescending. This FISH! principle invites us to advise our students in a positive manner and leave them feeling valued and capable of accomplishing their goals." As she points out, an effective advisor can "make a student's day" by "offering encouragement, praise, honest appreciation, and setting high standards."

Also, while the student is speaking, maintain eye contact with the student; do not look at your computer screen or out the window. Take notes as appropriate. After the student finishes describing his or her issues, briefly summarize the substance of the student's concerns to make sure that you understand them accurately. Once it is clear that you understand the student's concerns, summarize the action items that you plan to take and the action items that you encourage or expect the advisee to take.

The delivery of advice can be as important as the advice itself. Effective advisors deliver advice in a nonjudgmental, sympathetic, and nonthreatening manner. Sometimes, an advisor must deliver information that the student will not want to hear. For example, you may have to tell a student that he or she has been denied admission to a graduate program or that he or she will have to take more time than expected to complete degree requirements. Be honest as you do so, and take time to explain to the student the reasoning underlying this information. A curt "your request has been denied" can be devastating to the student; if you take time to explain to the student why the university has denied the request, the student is far more likely to understand your reasoning and is much more likely to come away from the meeting with a positive feeling.

Many advisors find it valuable to block off time specifically for advisor–advisee relationships. These meetings do not have to take place in your office; for example, meeting at a local coffeehouse is often an effective way for faculty members and students to communicate and exchange ideas with one another in a more informal setting.

Recognize That Each Student Is an Individual

Every student has a particular set of life circumstances, so take time to recognize the uniqueness of each student. The more an advisor interacts with the advisees, the more he or she will come to realize how different the students really are, regardless of what surface qualities may seem to link them together, such as age, gender, and major. This, of course, will lead to the giving of more tailored advice, which will help them even more in their life planning.

Experts on education have long recognized that students have different learning styles. Each student grasps concepts presented in class in a different way. Similarly, each student responds to advice, and to the advising process, in a different way. Do not assume that advice given to one student should be the same advice offered to another student with a similar set of issues. As an illustration, you may feel strongly that students should get some work experience before beginning graduate school or that every student should take a course in geographic information systems (GIS) to make him or her more marketable. While advice of this sort may be valid as a general principle, there may be situations in which it does not fit a specific student.

Avoid Stereotyping

A corollary to recognizing each student as an individual is the need to avoid making assumptions about individuals based on group identities and external characteristics. Most academics are aware of the need to avoid stereotypes associated with race and gender. Yet, we often invoke other stereotypes that are much more subtle but still potentially damaging to students. Do not assume that all scholarship athletes are "dumb jocks," that all students who are active churchgoers are narrow-minded and bigoted, or that all students from rural areas are slow to grasp new ideas.

Along the same lines, do not expect students to represent groups of which they are members. It is not reasonable to ask a female student—inside or outside of class—to explain how all women feel about a certain issue. Similarly, it is not appropriate to expect an African-American student to explain how the African-American community feels about a particular event or circumstance. Individual students should not be expected to represent their genders, ethnic groups, religions, or other groups of people to which they belong.

Put Yourself in the Student's Position

Similarly, it is easy to fall into the trap of assuming that what worked for you also works for your advisee. Suppose you took a certain sequence of courses in graduate school and found them valuable. Does this mean that your student should necessarily take the same courses? You and your partner may have worked out an effective balance between your professional activities and other parts of your life. Does that mean that your student must achieve the same balance? Advisors should draw on their own experiences; however, they need to be careful not to justify advice solely on the grounds that "this is what worked best for me."

Since students will be coming to you from all walks of life, it is important to be sensitive to individual needs and to anticipate how each student may deal with certain situations. Fleming et al. (2005) point out that "Advisers who understand students' learning styles and needs and who recognize the influence of the college's physical setting, the classroom setting, and peers on the academic and social development of advisees can provide the support that students need to reach their goals."

Recognize also that students are learning. As part of the learning process they will make mistakes. Students benefit from learning from their mistakes, so do what you can to help students learn to solve some problems on their own.

Know University and Community Resources

Suppose a student came into your office complaining of headaches, stomach aches, and other ailments and pains. Would you prescribe medical treatment? Or would you advise the student to obtain appropriate medical treatment from a physician?

Effective advisors know their limits, and they recognize that they cannot solve every problem themselves. NACADA's Core Value 2 states that "Advisors are responsible for involving others, when appropriate, in the advising process" (NACADA 2004). In many cases, effective advising includes referring students to resources in the university or in the community; in other cases, effective advising involves encouraging students to resolve issues on their own. Few faculty members and graduate students in the sciences or humanities have formal training in medicine or law. Advisors should be aware of and able to contact physicians, counseling centers, legal clinics, and other resources that can help students with these sorts of problems.

Occasionally, students exhibit dramatic changes in behavior that could be symptomatic of major and even life-threatening problems. Be ready to intervene if necessary. If in your interaction with a student you see clear evidence of depression, domestic violence, or drug abuse, be proactive. Call a medical professional or a counseling center to set up an appointment for the student. If possible, accompany the student to make sure that he or she keeps the appointment. Alert university officials to the problem, if appropriate.

APPLICATION OF THESE PRINCIPLES TO THE GEOGRAPHICAL SCIENCES

The principles described above are well established in the academic advising literature and can be applied to students in many disciplines. However, advising students in geography and related fields involves a unique set of challenges associated with the nature of the discipline itself.

What makes advising in the geographical sciences unique? In many disciplines, curricula are geared toward preparing students for a fairly narrow range of career opportunities. In many cases, this preparation is linked to establishing formal credentials associated with entry into a profession. For example, an important component of law school is preparation for the bar examination, which a fledgling lawyer must pass in order to be allowed to practice law. Prospective physicians must not only complete medical school but must also complete internships, residencies, and other activities leading to certification before they can practice medicine. Degree programs in accounting are designed to prepare students to pass examinations associated with certification as a public accountant. The presumption is that students enrolled in these programs are preparing to obtain this certification, and curricular requirements and advising are undertaken accordingly.

In contrast to education in these and other professions, education in fields such as geography is relatively independent of formal credentials. As geography professors are well aware, geography is a very broad discipline. At all educational levels, students with degrees in geography have access to a broad range of professional opportunities. Many students decide to major in geography because of its flexibility, breadth, and integrative nature. The flexibility and breadth of geography makes it difficult, however, for advisors of geography students to make blanket assumptions about students' educational and life goals. Advisors should not assume that each student is in a geography program with the intent of preparing for a specific occupational goal.

As an example, many social and environmental science programs now emphasize training in GIS. While such training is valuable, in some institutions GIS education is presented and advocated by advisors as a way to ensure that the student will "get a job." But what types of jobs are such students being prepared for? Some GIS jobs are challenging positions requiring considerable preparation in geography and spatial analysis, whereas others are technician positions that require different sets of skills. This example reinforces the point that each student is an individual and that the advisor should not assume that what will work for some students will work as well for others.

ACKNOWLEDGMENTS

Portions of this paper were presented to the participants in the Geography Faculty Development Alliance workshop in Boulder, Colorado, in June 2006 and to the 103rd annual meeting of the Association of American Geographers in San Francisco, California, in April 2007. The helpful comments of Susan Hardwick, Emily Murai, Ken Foote, and five anonymous referees on earlier drafts of this paper are gratefully acknowledged.

Additional Resources

Many articles have been written, and presentations given, on the practices of advising, and it is here that we list two of the most comprehensive sources for academic advisors.

The **National Academic Advising Association (NACADA)** is an organization that serves academic advisors of higher education institutions. NACADA provides many publications and opportunities for discussion, all of which can be accessed on the web at http://www.nacada.ksu.edu. Below, we list specific links to certain parts of their web site.
- *The NACADA Journal*, a biannual refereed journal, http://www.nacada.ksu.edu/Journal/
- *Academic Advising Today*, NACADA's quarterly electronic publication, http://www.nacada.ksu.edu/AAT/

- *Clearinghouse of Academic Advising Resources,* a central collection point within the web site for member-suggested resources for academic advisors, http://www.nacada.ksu.edu/Clearinghouse/overview.htm#mission

The Mentor: An Academic Advising Journal, which is published by Penn State's Center for Excellence in Academic Advising, is an electronic publication that supports effective academic advising. The editors welcome contributions of all sorts, especially short articles concerning "examples of innovative advising programs, summaries of conference presentations, descriptions of exemplary practices in advising, letters to the editor, and other concise forms of writing related to advising." The journal can be accessed at http://www.psu.edu/dus/mentor/, and the web site itself offers many other resources, such as a forum which changes topics monthly, and "The Muse," a section of the site devoted to poems, short stories, and cartoons as they relate to advising.

References

Appleby, D. 2001. *The teaching-advising connection: How do faculty perceive their roles as academic advisors?* http://www.psu.edu/dus/mentor/appleby1.htm (last accessed 26 September 2007).

Chickering, A. W. 1994. Empowering lifelong self-development. *NACADA Journal* 14 (2):50–53.

Clemson University, Academic Advising. 2001. Academic advising goals and objectives. http://www.clemson.edu/advising/goals.htm (last accessed 17 September 2007).

Creamer, D. G., and E. G. Creamer. 1994. Practicing developmental advising: Theoretical contexts and functional applications. *NACADA Journal* 14 (2):7–24.

De Sousa, D. J. 2005. *Promoting student success: What advisers can do.* Bloomington, IN: Indiana University, Center for Post Secondary Research. Occasional Paper #11. http://nsse.iub.edu/institute/documents/briefs/DEEP%20Practice%20Brief%2011%20What%20Advisors%20Can%20Do.pdf (last accessed 17 September 2007).

Fleming, W. J., B. K. Howard, E. Perkins, and M. Pesta. 2005. The college environment: Factors influencing student transition and their impact on academic advising. *The Mentor: An Academic Advising Journal* 7 (3). http://www.psu.edu/dus/mentor/(last accessed 17 September 2007).

Friedman, T. L. 2005. *The world is flat: A brief history of the twenty-first century.* New York: Farrar, Straus, and Giroux.

Gregory, C. 2002. How is academic advising different from teaching, personal counseling, and career counseling? *The Mentor: An Academic Advising Journal* 4 (1). http://www.psu.edu/dus/mentor/ (last accessed 17 September 2007).

Hardwick, S. W. 2005. Mentoring early career faculty in geography. *The Professional Geographer* 57 (1):21–27.

Hardwick, S. W., and F. Shelley. 2007. Effective mentoring in geography. Washington, DC: Association of American Geographers, Healthy Departments Initiative. http://www.aag.org/healthydepartments/healthy_resources.cfm (last accessed 17 September 2007).

Hennessy, V. 2004. Let's go fishing: Embracing students in academic advising. *The Mentor: An Academic Advising Journal* 6 (1). http://www.psu.edu/dus/mentor/ (last accessed 17 September 2007).

Light, R. J. 2001. *Making the most of college: Students speak their minds.* Cambridge, MA: Harvard University Press.

Lundin, S., H. Paul, and J. Christensen. 2000. *FISH!: A remarkable way to boost morale and improve results.* New York: Hyperion.

Moss, P., A. Cravey, J. Hyndman, K. K. Hirschboeck, and M. Masucci. 1999. Towards mentoring as feminist praxis: Strategies for ourselves and others. *Journal of Geography in Higher Education* 23 (3):413–27.

NACADA. 2004. NACADA statement of core values of academic advising. http://www.nacada.ksu.edu/Clearinghouse/AdvisingIssues/Core-Values.htm (last accessed 17 September 2007).

Pennsylvania State University, Center for Excellence in Academic Advising. 2003. Advising as teaching. http://www.psu.edu/dus/cfe/rolteach.htm (last accessed 17 September 2007).

CHAPTER 9

Ethical Teaching in Practice

James Ketchum

Professional development during graduate education often lacks an explicit consideration of the ethical dimensions of teaching—that is, those codes of behavior that support professional standards of competency, encourage fairness and respect for one's colleagues and students, oblige us to teach in a manner consistent with institutional goals and values, and lead us toward a personal dedication for continued improvement and development in the area of teaching. Whereas some professors hold deeply felt ethical understandings of teaching, many others do not largely because they have never been *asked* to think of teaching in terms of its ethical components. As a consequence, many aspiring academics begin their careers "with very little capacity to analyze moral problems" (Smith 1995, 272), and thus often find it difficult to manage issues ranging from plagiarism and discriminatory behavior to explaining the ethical underpinnings of disciplinary subject matter.

As I hope to show in this chapter, ethics underpin every activity we undertake as a college or university instructor, from designing course syllabi to our policies for student assessment and classroom management. In approaching the subject, my aim is to prompt critical reflection about the aspects of your teaching that embody ethical principles, using examples from the literature as a starting point for this dialogue. After reading this chapter, I encourage you to explore the case studies presented in Activity 9.1, which is obtainable on this book's web site. The case studies are intended to help you identify the various ways that ethical situations and dilemmas arise in college classrooms and to show how, by adopting a set of ethical principles outlined below, you will be better prepared to manage similar incidents in your own career.

Having a dialogue about ethical teaching in practice does not imply a sermon about "right and wrong" behaviors, yet at the same time the realities of the contemporary academy require us to avoid the other extreme of thought—that "everything is relative" when it comes to setting professional standards for teaching. It is true that many academic professionals value personal autonomy, especially with regard to teaching. This is easy to understand because our teaching is often not subject to the same institutional scrutiny as our research (as is the case with Institutional Review Boards that review human subjects protections, a topic further explored in Chapter 12 "Ethical Research in Practice" by Iain Hay and Mark Israel). Indeed, the typical college-level instructor works without much if any direct supervision, aside perhaps from occasional classroom observations by colleagues. But it would be a mistake to conclude from this that the decisions you make as a teacher will not translate across your other professional identities: as a researcher, as a colleague and citizen in your department, as an adviser to your own graduate students, and as a public servant. We must remember that not only do students put their faith in the integrity of the professor, but so does one's academic discipline, the academic department or program in which one works, the university the professor represents, and even society as a whole. Your employer, your students, and your colleagues will expect that you will conduct yourself professionally according to certain ethical codes of behavior. Some academic institutions and departments define these expectations in faculty handbooks, whereas others do not.

What I argue, foremost, in this chapter is that ethical practice in teaching is strongly connected to effective practice as an academic professional more generally, and that by committing yourself to developing an "ethical imagination" (cf. Proctor 1998) right at the beginning of your career, you will strengthen your own instructional development and contribute to the overall quality of professionalism in your department and institution.

IMAGINING YOUR TEACHING IN ETHICAL TERMS

Perhaps the most important reason to frame the consideration of teaching practices in terms of ethics is simply this: impressionable students typically place a great deal of trust in the integrity of the professor, placing faith in his or her professionalism, judgment, and training. The professor determines the course content; creates, manages, and maintains the environment of the classroom; preserves a sense of fairness and justice within that environment; negotiates disputatious points of view between students; and creates instruments for assessing learning, among other responsibilities. In the volume *Ethical Dimensions of College and University Teaching: Understanding and Honoring the Special Relationship Between Teachers and Students*, Fisch (1996) presents several principles that can help you strengthen the ethical foundations of your courses. I list these in Figure 9.1 and also note how they collectively can benefit your overall professional development as an academic.

Later, as you work through the cases in Activity 9.1, think about how these principles might guide your decisions and help you manage similar situations that may arise in your own class.

FIGURE 9.1 Nine principles for ethical instruction (Fisch 1996), with annotations by the author.

1. **Content competence:** This principle asserts that, as a professional responsibility, teaching requires strong knowledge of disciplinary subject matter and awareness of new discoveries and ways of thinking in a field. Because you cannot always expect to teach courses in your area of research specialization, you may find it necessary from time to time to beef up or refresh your knowledge of other subject areas.

2. **Pedagogical competence:** Building on the first principle, teaching also requires an appreciation of learning theory and the ability to call upon a diverse repertoire of teaching strategies. The chapters by Dee Fink and Melissa Ganus ("Designing Significant Learning Experiences") and Eric Fournier ("Active Learning") can help you make these connections.

3. **Dealing with sensitive topics:** Disciplines such as geography often deal with controversial issues. Realizing that students carry different sets of values and experiences, and thus will react differently to particular issues, this principle reminds us that it is incumbent upon the instructor to create a classroom climate that respects difference while allowing for open, honest, and positive dialogue. The chapter by Minelle Mahtani ("Teaching Diverse Students: Teaching for Inclusion") addresses strategies for creating such a climate.

4. **Student development:** Perhaps the fundamental aim of teaching is to foster the intellectual growth of learners. As noted above, a welcoming and respectful classroom environment is one of the ways you can enhance the likelihood that your own teaching will enhance student learning. Similarly, you can contribute to student development by serving as an undergraduate or graduate advisor (Fred Shelley and Adrienne Proffer provide useful tips in their chapter "Advising Students").

5. **Dual relationships with students:** As a corollary to the fourth principle, you must take pains to avoid conflicts of interest (such as romantic involvements) and other actions that can lead to favoritism toward certain students and detract from the development of others.

(Continued)

FIGURE 9.1 Continued

6. *Confidentiality:* This principle increasingly governs the administrative aspects of faculty lives, requiring you to safeguard student grades, attendance records, and private communications. However, occasions may arise when this information can be released per the consent of students; in other cases, you may become privy to information that compels you to take action in order to protect the rights of others. For a related discussion of confidentiality issues, see Iain Hay and Mark Israel's chapter "Ethical Practice in Research."

7. *Respect for colleagues:* Although teaching has been traditionally viewed as a "private" practice, you should not fear inviting constructive critique or offering to help other faculty who are just getting started in the classroom. Indeed, as Nellis and Roberts note in their chapter "Developing Collegial Relationships," asking others for suggestions and advice about teaching—and being willing to mentor others—can lead to healthy dialogue and strengthen working relationships within academic departments.

8. *Valid assessment of students:* One of the dominant trends in higher education is the increasing pressure on academic institutions to be held accountable for student learning outcomes. Inevitably, this responsibility will be felt by the faculty teaching the students. But assessment is not an administrative chore; it is a crucial component of the teaching and learning process. As Dee Fink and Melissa Ganus point out in their chapter "Designing Significant Learning Experiences," there are a number of assessment strategies that accurately measure what students learn and are able to do as a result of instruction.

9. *Respect for institution:* This final principle does not mean that you, as an employee, should steer away from questioning the academic policies of your institution. But it does aim to promote awareness among faculty of the overall educational goals, policies, and standards of the institution. By designing your courses in a manner that complements the mission of your department and institution, you will contribute to the overall quality of the learning environment on campus. When colleagues view your professional activities as serving the interests of the institution and its stakeholders, this can only help position you more favorably for success at the time of tenure review. Susan Roberts further explains these relationships in her chapter "Succeeding at Tenure and Beyond."

In contrast to thinking about teaching in terms of "right versus wrong" things to do, Martin (1995) suggests a more insightful approach would be to consider ethical issues in relation to the varied contexts in which professors teach students. Although the principles outlined in Figure 9.1 are transferable across all types of academic institutions, ethical teaching requires us to imagine ourselves as ethical actors able to consider the needs, motivations, and values of diverse groups of students, colleagues, and campus administrators who bring different sets of experiences and expectations to the profession. The case studies provided in the chapter activity thus may benefit you the most if you discuss them with colleagues in your department. Because all ethical situations must be interpreted by individuals from their own perspectives, a collective dialogue will help reveal commonalities and differences in how ethical standards can be attained for the common good. Such discussions, moreover, can inspire critical thought and ethical imaginations by helping you understand that all situations involving ethical questions are interpretable—perhaps not limitlessly so, but certainly subject to a variety of valid interpretations, nonetheless.

Imagination is also valuable because while it might seem possible that a normative "code" of ethics could be applied to guide our teaching practices, the complexities of real-life circumstances often make that difficult. In fact, many ethical situations "from which incontrovertible moral and ethical guidance might be drawn often suggest irreconcilably different behaviors" (Hay and Foley 1998, 171). To illustrate: when, if ever, should a late paper be accepted without penalty? What if a student turning in a late paper confesses experiencing relationship difficulties with his or her spouse? Or what if the reason given was a night of binge drinking, but by penalizing the student with a low score, he or she would receive a failing grade and lose a financially essential scholarship? Where, exactly, is the ethical line drawn? I could continue to spin variations on this scenario, and in all likelihood you will face these sorts of dilemmas throughout your career. In this particular case, having an ethical framework early in your career would give you the foresight to incorporate clear, unambiguous language about class policy regarding late papers, and save you from making on-the-fly judgments that can quickly produce additional difficulties.

Another aspect of ethical thinking emerges from the fact that we usually think of ethics as involving a specific incident, such as a dispute over grades. Many of us think of an ethical problem as a special case or some circumstance that interrupts the flow of our everyday routine, has to be dealt with, but then goes away and leaves us to return to our research and writing. However, what we often fail to recognize is that ethical considerations are part of almost everything we do, including the day-to-day practices we follow in our teaching. Sometimes a special case arises, but ethics more often involves simple day-to-day habits. We know we must adhere to certain ethical habits in the practice of research, but ethics can and should be applied to the habitual behaviors of teaching, as the principles in Figure 9.1 make clear.

In fact, by viewing the academic profession as an ethical enterprise we can make our jobs easier by establishing frameworks for action which will often prevent unusual conflicts from arising in the first place.

For example, plagiarism by students has recently been on the rise, partly due to Internet web sites that offer many thousands of essays to students at the touch of a button (for a price), an option that may become increasingly attractive to a student on a late night before a due date. Plagiarism, although now more easily achieved, may also be more easily detected, thanks in part to readily available software programs made specifically for professors. We must also recognize that different institutions have differing regulations on plagiarism. Some go by the honor code; others are more specific and formal.

Detecting plagiarism is not necessarily the real problem, however. The more sensitive difficulty is in dealing with it, and this is where our ethical imagination may be of service. Specifically, it can help us recognize that every case of plagiarism is different, and although wrong, the real ethical issue involves how we might best serve the student and all those groups to which we are responsible—the university, discipline, department, and society—by our response. While plagiarism is a breach of ethical conduct by the student, we must also understand that our response to that plagiarism carries its own ethical considerations. We should of course make clear the institutional policies governing plagiarism and work toward producing the circumstances that will prevent it from taking place. But if we do not take into account the ways that students may legitimately be struggling to understand the problem, if we fail to really explain why it is so important to avoid, and if we neglect the heavy emphasis we should place on its avoidance at the beginning of the course, are we really doing a service to the student, to our university or college, to our discipline, and to our department by condemning the student?

In the same vein, what is considered cheating—or example, on exams—differs from institution to institution. Therefore, as one of the principles in Figure 9.1 stresses, you should recognize the standards and policies of your institution, and orient your teaching in a manner consistent with these codes so that students are not given conflicting messages. The teacher needs to make sure that the student has a very clear idea of what is and is not expected. You may want to consider this when thinking through the case studies in Activity 9.1.

COMMITTING TO ETHICAL TEACHING

In order to teach ethically we must commit ourselves to the job, but we often neglect to consider what this entails. If we think about the time we spent as students, each of us can remember at least one professor who seemed to be simply going through the motions of teaching without much interest or passion. Perhaps interested in teaching someday ourselves, we may have even told ourselves that we would never do the same thing, and that instead we would always teach the right way by being prepared, passionate, and even tireless in our efforts to engage students meaningfully each and every day.

Unfortunately, what we experience in graduate school or the first few years of an academic appointment may suggest that the reality of teaching can be very different, especially today. With increasing pressures to publish more and more articles, to obtain outside funding through grants, to serve on committees, to increase our teaching effectiveness by attending workshops, and to reach outside the university to engage the community by applying our research to concrete local problems, even the most idealistic of university and college teachers may have found themselves putting less and less time into their classes despite their best intentions, and even, on occasion, cutting corners to get through the next day. Failing to update lectures, rethink exams, provide useful written feedback on papers, or even attend our own office hours are just some of the ways we sometimes fail to do what we know we should do. But as Ken Foote makes clear in Chapter 1 on time management, achieving balance and moderation in our professional lives may be a struggle, yet we can do it, and we can also help others.

We have seen many professors teaching in college classrooms, but how does each of us, as individuals, adopt the expectations of various groups and individuals and productively embody them? How will we each find our voice as a teacher? For some this is easier than others, but one of the most important factors in learning to teach productively (and ethically) is to recognize that others may see you differently than you see yourself. Ethical issues often arise when a professor fails to recognize this difference. For example, sometimes, in seeking to relate to students, a professor will talk too much about his or her own personal life in the classroom or will interject too much of personal opinion into the act of teaching the course. But how do we define "too much"? And how could this be an ethical problem?

Imagine that a professor, Dr. Martin, comes into class and tells a story about his personal life. Always feeling a little uncomfortable by being placed in a position of authority in front of a room of relative strangers, perhaps the professor allays that discomfort by making it clear that he is just a nice, regular guy who does things everybody else does when not put in the position of standing in front of a blackboard, teaching. The ethical problem might be that this professor either fails to recognize or take seriously his professional role as teacher, and in doing so he fails to uphold his commitment to the institutions that sanctioned his work, including the university, the department, and his academic discipline. Also he fails to respect the expectations of his students, all of whom most likely feel unwilling to suggest that the teacher should just get on with teaching the class. Uncomfortable with his role as a teacher and the expectations and responsibilities that role entails, he fails to act according to his unwritten code of professional ethics, which at the very least should involve respecting the needs and expectations of paying students within the allotted time frame. Such professional behaviors should also extend to simple, everyday acts such as keeping one's office hours, beginning and ending class on time, respecting the schedule on the syllabus, and avoiding canceling class if at all possible.

UNDERSTANDING ETHICAL TEACHING IN THE CONTEXT OF A DISCIPLINE

Disciplines themselves provide a context for weighing the implications of teaching practices. For instance, as early as 1885 in the discipline of geography, Peter Kropotkin wrote of the special relevance of morality and ethics in geographical content. For Kropotkin, geography teachers had a responsibility through their teaching to counteract the "hostile influences" and "absurd prejudices" at work in society (Kropotkin 1996, 141). Undoubtedly, this line of thinking remains one of the primary ways that contemporary geographers think of their work as teachers. Often charged with teaching about the globe and its peoples in the context of a synthesizing, holistic worldview, many geographers feel a moral obligation to teach students about that which the world's peoples share in common, perhaps to counteract divisive forces that can lead to tense relations between ethnic groups and nations (Merrett 2000).

Recall that Fisch's first principle of ethical instruction requires instructors to remain aware of new discoveries and new ways of thinking in a discipline. Certainly, we should not stick our heads in the sand like ostriches and fail to even *try* to understand the ways that our teaching might help to shape students' knowledge in *unintended* ways. In moving toward both content and pedagogical competence, we must understand that the truly ethical approach in this case is to try and appreciate the criticisms and perspectives of others, not least those of students who differ from ourselves by, for example, ethnicity, class, or cultural background (see Chapter 10 by Minelle Mahtani).

Consider, finally, an example of an ethical issue in the context of geographic information systems (GIS). Many college professors use spatial data sets and geographic information technology to teach about topics as diverse as crime, public health, environmental pollution, and transportation. With spatial data, it is fairly easy to reveal information that could compromise personal privacy. If working on a project examining geographical patterns of crime, one might easily provide the address of the individual crimes in order to chart them on a map. But it is a short step to connect an address with a name, and many victims of crime would prefer keeping that information private. Here, a confidentiality issue emerges that is quite similar to those affecting research practice (for examples, see Chapter 12 by Iain Hay and Mark Israel).

There are many other ways that ethical issues can arise in ways specific to teaching geography, but the important point here is to recognize the overall importance of rethinking teaching in terms of its ethical components. Disciplines such as geography often deal with sensitive or controversial issues, but the real key is to develop a sensitivity to the various experiences, values, and circumstances that may inflect the opinions and actions of everyone concerned in the discussions in which we find ourselves.

Additional Resources

Hay, I. 1998. From code to conduct: Professional ethics in New Zealand geography. *New Zealand Geographer* 54 (2):21–27.

Israel, M., and I. Hay. 2006. *Research ethics for social scientists*. Thousand Oaks, CA and London: Sage.

MacKinnon, B. 2006. *Ethics: Theory and contemporary issues*. New York: Wadsworth.

Smith, D. 2001. Geography and ethics: Progress, or more of the same? *Progress in Human Geography* 25 (2):261–68.

Valentine, G. 2005. Geography and ethics: Moral geographies? Ethical commitment in research and teaching. *Progress in Human Geography* 29 (4):483–87.

References

Fisch, L., ed. 1996. *Ethical dimensions of college and university teaching: Understanding and honoring the special relationship between teachers and students*. San Francisco: Jossey-Bass.

Hay, I., and P. Foley. 1998. Ethics, geography and responsible citizenship. *Journal of Geography in Higher Education* 22 (2):169–83.

Kropotkin, P. 1996. What geography ought to be. In *Human geography: An essential anthology*, eds. J. Agnew, D. Livingstone, and A. Rogers, 139–54. Oxford, U.K.: Blackwell.

Martin, M. 1995. *Everyday morality: An introduction to applied ethics*. New York: Wadsworth.

Merrett, C. 2000. Teaching social justice: Reviving geography's neglected tradition. *Journal of Geography* 99:207–18.

Proctor, J. 1998. Ethics in geography: Giving moral form to the geographical imagination. *Area* 30 (1):8–18.

Smith, D. 1995. Moral teaching in geography. *Journal of Geography in Higher Education* 19 (3):271–83.

CHAPTER 10

Teaching Diverse Students: Teaching for Inclusion

Minelle Mahtani

Let us begin with a scenario. You have just started your first faculty appointment. You have been assigned to teach the introductory course and today's topic is religion and identity. Since you are still new to teaching, you are not surprised to have butterflies in your stomach. However, you feel more upbeat than usual because you spent hours prepping. You step up to the podium, introduce the day's topic, and everything seems to be going well, until you mention Islam and postcolonialism. Your presentation is disrupted by a clearly audible comment from a student: "I'm not a racist, but let's get real here—Islam is a fundamentally violent religion." Clearly, many students heard the remark, so you are about to acknowledge it by saying "I know some people have very negative views about Islam, but it is unfair to homogenize an entire religion." Before you can comment, another student interrupts, "Did you hear that? It's the most racist thing I have ever heard! I mean, I am Muslim! What she is saying is hate speech!" The class starts to buzz.

The moment is loaded; it is a "hot-button" issue. Not only has the first student made a grossly unfair generalization, but she has also offended another student. This situation could trigger divisions in the classroom that could hinder further discussion and potentially alienate both of the students and some of their classmates for the duration of the semester, or it could be an opportunity for learning to promote respectful discussion and curiosity. What do you do?

While the above example is fabricated, it is not dissimilar from situations many professors have faced in the classroom. As university educators

121

who teach sensitive topics in the social sciences and the humanities, we often encounter moments that are heavy with tension. As Kobayashi, who teaches a course on race and racism, explains, "I sometimes feel as if I am carrying a bomb into class, and if I am unsuccessful in establishing the right degree of comfort (or discomfort) it will explode with irreversible results" (Kobayashi 1999, 179). Such moments, however, offer an opportunity for growth, change, and development for our students and for us as educators.

In fields such as geography that marry the social and environmental sciences, challenging situations can also arise when we engage our students in short or extended periods in the field outside the classroom. How do we design those experiences to include students who have work and family responsibilities or physical disabilities? What issues arise when we ask students to map or photograph urban neighborhoods where they may feel insecure or, alternatively, sensitive when their peers pass judgments about class or ethnic landscapes with which they identify (Nairn 2003)? In other words, what does it mean to truly support diversity—of all kinds and in all the varied learning environments where you may someday teach?

This chapter explores the many faces of diversity in the university environment. It pays particular attention to how we can create a culture that welcomes and supports a range of voices and identities. It asks how an aspiring academic can make positive change through pedagogical practices and curriculum design and by questioning the assumptions and preconceptions we and our students sometimes carry into learning environments. I will begin by defining what I mean by diversity. I then offer on this book's web site some activities you can use to promote an environment that supports diversity. I hope that you will learn strategies and ideas that will set in motion a continued commitment to diversity concerns in teaching, research, and service.

WHAT IS DIVERSITY AND WHY DOES IT MATTER?

> Diversity is the empirical reality that we must explore for the sake of our scholarship. (Lee 1997, 263)

As educators, it is important for us to keep in mind the many forms of diversity our students bring to the university environment. Regardless of whether you are at an urban or a rural institution, a small liberal arts college, or a massive research university with four campuses, your student body will always be diverse in some way. Paying attention to diversity in its myriad forms means remembering that diversity is not only about race, class, or gender—although it is, of course, always about these dimensions. It is our responsibility not only to deconstruct the concept of difference as it relates to these crucial identifications but also to keep in mind differences among students such as physical and mental disabilities, part-time and full-time enrollment, age, nationality, sexuality, religion, culture, language, and regional origins, among many other markers of identity (Brown 2004). It is also important to think about how different axes of identity intersect for students. What might the challenges be for a student who is not only a quadriplegic but who may speak English as a second language? It goes without saying, too, that differences are not always visible. Learning disabilities such as dyslexia may not be immediately noticeable, and you cannot assume that your institution will provide you with advance notification of the presence of students with learning disabilities and how they can be accommodated.

Although the benefits of teaching through diversity may seem evident, to my surprise, I have often met resistance and skepticism about the topic from early career faculty. "Diversity matters—sure," I am told. "But I am too busy designing my syllabus and attending faculty meetings to really pay attention to it—maybe next year once I get my feet wet." Herein lies the first problem: when diversity is considered as an afterthought to curriculum design and faculty commitments. For if we, as the leaders within the academic community, hesitate to focus our analyses through the lens of diversity, we necessarily teach our students (and fellow faculty) by example that diversity is indeed secondary. Without a personal set of reasons and understanding of why diversity matters, you will not be able to skillfully or authentically navigate a discussion on diversity issues in the classroom and will become lost or uncomfortable during a "hot-button" moment.

Diversity is important to address now—not later—because doing so can help you realize the educational goals you have set for your classes. The students we teach and the worlds in which they live are becoming increasingly diverse; equipping them with the tools they need to understand these changes is one of our important tasks as educators (Banks 1994). A wealth of research in this area indicates how vital it is for us to understand, acknowledge, and support students who both possess and express diverse perspectives. Not only do we recognize more diversity within native-born student populations, but in the last twenty-five years international student mobility has also risen dramatically to increase the range of cultural assumptions and values we will encounter among students as well as the presence of nonnative speakers of English (Brunch and Barty 1998).

Numerous educators have lamented the lack of equitable representation of diverse groups in textbooks and in the college curriculum more broadly. Absences are evident in whose voices (which authors) are cited, whose lives are portrayed in textbooks, articles, and visual materials, and in pervasive stereotyping, for example, by gender, ethnicity, culture, or world region. Are all immigrants and "heads of household" male? Are women generally shown in domestic roles, but men never or rarely? Does "race" only enter as a topic when urban problems are discussed? Are African examples confined to themes of disease, environmental deterioration, or poverty? Furthermore, the strengths and skills of diverse students are often not acknowledged or addressed. Teachers do not make it a habit to discuss the perceptions that students have of themselves and each other and how these can influence the learning process (Sleeter 1989). Unfortunately, teaching diverse students has been treated as a sideline or an add-on, primarily because professors are rarely trained or encouraged to think about the issues facing marginalized students (Haigh 2002). Yet, as James Ketchum points out in Chapter 9, ethical practice as teachers requires that we recognize and respect differences.

HOW CAN YOU SUPPORT DIVERSITY?

There are many ways to support diversity in the university environment. A critical starting point is to question your personal assumptions, judgments, and understandings of difference and diversity. An effective teacher committed to cultivating a classroom environment that supports diversity adopts a nonjudgmental view of difference to the extent that differences of opinion, belief, and orientation are welcomed and invited as opportunities for robust learning.

It is also crucial to recognize the talents and abilities of all your students and to see them all as capable learners. Research has indicated that teachers who perceive particular students as being deficient or incapable are likely to expect less of them (Nieto 1996). The ways in which gender issues create "chilly" classrooms have been identified in numerous studies, particularly in relation to conditions that make women students reluctant to speak (Nairn 1997). For example, researchers have documented the silencing effects of posing analytical versus factual kinds of questions to men more often than to women (Sandler 2007).

HOW CAN BIASES BE RECOGNIZED?

Before we begin to address the issue of sensitivity and diversity, it is imperative to explore the biases and assumptions we bring into the college classroom. You may not think you have any, but we all do, and they influence our teaching, as well as our day-to-day choices in the classroom environment. What biases might you have absorbed in your life? Are you unintentionally sending different messages to different groups? Figures 10.1 and 10.2 pose some questions for you to

FIGURE 10.1 Examining your own biases.

- How do your personal experiences inform the way you interact with individuals whose background is different from yours?
- Am I comfortable around students of color? Students with a disability? ESL students? Do I find myself acting in a different way around them?
- Am I uncomfortable when students become emotional in my classroom?
- If an issue about race/gender/class/nationality/religion/disability, etc., comes up, do I assume that a student of that background knows something about it?

FIGURE 10.2 Reflecting on your teaching practices.

- Do I call on traditionally marginalized students as often as I do others?
- Do I ask students to "speak for" the perspectives of the group (e.g., ethnic, gender, national origin) with which I associate them?
- Do I believe there is one way of arguing or discussing ideas in class?
- Does it sound like an "exclusive club" when I tell stories about conferences and other events in our discipline?
- Do I tend to shelve or "make time later" for minority points of view?
- Do I bring in diverse opinions of people from marginalized groups in my classroom?
- Do I include readings in my syllabus by or about people from marginalized groups?

think about—they are aimed at prompting you to be more self-reflexive about your beliefs and behaviors when dealing with different kinds of students.

It is worthwhile to examine your preconceived beliefs and pay particular attention to moments that create personal discomfort. How can you work to ensure that you are not avoiding challenges because you fear feeling uncomfortable? Villegas and Lucas (2007) suggest that seeing traditionally marginalized students in more positive ways is productively linked with enhancing intellectual rigor in the curriculum, setting higher performance standards, and effectively building on the skills and talents these students bring to the classroom. Thus, it is important not only to look at our own biases but to also examine how we might build bridges across the curriculum to connect with students with diverse backgrounds.

THE IMPORTANCE OF KNOWING WHERE YOU ARE

While knowing yourself is the first step, the second step is to learn more about the environment and place you are in. Most of us teach in radically different places from those where we grew up or earned our degrees. Discovering more about the geography of where you teach can prove very helpful in understanding the diverse range of students you will encounter. What is the socioeconomic background of the majority of your students? Does your college have a predominantly Latino population? Are you teaching at a commuter school where many students hold full-time jobs? Speaking to the diversity coordinator at your institution will provide you with very useful information about the makeup of those traditionally marginalized in the student body. For example, I have taught at large urban universities where the majority of students speak English as a second language.

In thinking through how I could encourage the maximum learning potential for these students, it was important for me to use simplified language, speak more slowly, use more visual cues, and integrate more hands-on activities in my lessons. This did not mean "dumbing down" my material—in fact, quite the opposite. This process also required me to ensure that my teaching style and assessment methods reflected a multidimensional approach that included specialized teaching techniques, and critiquing my own biases or stereotypes about particular populations (see also Scheyvens, Wild, and Overton 2003 who examine issues facing international students pursuing postgraduate degrees in geography).

When considering how to make your classes supportive of diversity, it is also important to recognize that what may "work" in one context may be problematic in another. Saunders (1999) offers a case study of markedly different responses when he showed students rap music videos to illustrate ways in which racism is portrayed, and also resisted, in popular culture. In his class at the University of Arizona, where most students were white, responses were usually positive, though he wondered if his teaching had any substantial effect on their thinking over the longer term. When he used the same material in a more diverse classroom in California, the student reactions were more varied and often emotionally charged. African-American students were divided. Some challenged him with stereotyping ghetto life, others (older African-American women) objected to what they felt was offensive language in rap music, and still others wanted to talk about the dangers of growing up in inner city neighborhoods. A group of white students were interested in gaining new perspectives, whereas perhaps half the class, especially Latinos and Asian-origin students, seemed less engaged. The situation prevented Saunders from getting to his main goal of discussing the politics of representation and of rap as a discourse challenging racism. But the experience confirmed his commitment to sustained discussions of race and racism.

A commitment to diversity means cultivating a commitment to teaching that takes into consideration the roles of culture and language in learning. Research on diverse learning practices has indicated that the way we learn best is through using prior knowledge and belief systems to make sense of the world (Villegas and Lucas 2007). It is important to build bridges between what it is that students already know and what they need to learn. It also means affirming their diverse identities in class. Students appreciate it when they can see themselves reflected in classes. Take time to find examples that are pertinent and resonate with students' lives and experiences. Encourage critical thinking in your classroom by questioning, analyzing, and deconstructing the assumptions inherent in textbooks and other educational media. When biases and stereotypes are found, I often ask students to consider how the content could be written or approached differently. This gives students the responsibility for their own learning as well as an opportunity to engage with the material on an entirely new level.

SUPPORTING DIVERSITY AND INCLUSIVENESS WITH YOUR SYLLABUS

One way to see how you score on diversity issues is by examining your syllabus. While at first glance it may seem to be addressing issues of diversity, asking some new questions about your syllabus may point to areas where it could be improved. Pull out the syllabus for a course you teach, and then evaluate it using the following questions.

Does the Syllabus Include Guidelines for Class Conduct?

You must set ground rules early in the course, stating what you consider to be appropriate and inappropriate behavior. Students take their cues from us; they model their behavior on how we act and respond. Your syllabus can include a statement about encouraging diverse perspectives and points of view in the classroom. Here is an example from one of my classes "Spaces of Multiraciality: Critical Mixed Race Theory":

> Because this course addresses many contemporary and controversial issues—discrimination and power, gender, language, racism, sexuality—I insist that each of us respect the thoughts and opinions of one another. Our tutorials and lectures are intellectual forums to explore cultural issues from a geographical perspective; everyone's voice and interpretations are welcomed. As the course instructor, I will provide you with a set of theoretical concepts, models and various interpretations of cultural conflicts. Using these models and concepts, you will be encouraged and challenged to develop your *own* interpretations of various cultural conflicts. In other words, your grounded opinions in no way whatsoever need parallel my own to succeed in this course. Freedom of thought expressed in a courteous manner is encouraged. If we all abide by a code of civility and mutual respect, we create a mature, safe and intellectually stimulating forum. Finally, if you have problems or questions with any aspect of the course, please raise the issue with me in class or during my office hours. I am approachable, and I welcome your questions and constructive criticisms.

A statement like this will help set the stage and parameters when "hot-button" topics do come up.

Is the Syllabus Sensitive to Terminology?

Is the language of your syllabus sensitive to nuances in terminology? Consider carefully names and terms that may be divisive if used thoughtlessly, especially those which may be challenged because their meanings are in flux or can change depending on where, when, and with whom they are used. For example, do you use Native American/American Indian/Indian/Indigenous;

disabled/handicapped; illegal/undocumented (in relation to immigrants); Mexican American/Chicano(a)/Hispanic/Latino(a)? Also think about words that might be considered sexist and reinforce gender stereotypes, for example, referring to women students as "girls" (but not to men students as "boys"). If you are uncertain about usage, and especially about local preferences, contact the diversity office on your campus.

What Voices and Groups Are Represented in the Reading Assignments?

It is worthwhile to take a closer look at the textbook you use to see if it simply "adds on" a section on diverse cultural identities (a practice not uncommon in second or later editions). Ironically, it may be instructionally valuable to include biased materials so that students can more easily identify the biases or, if they fail to do so, to address the examples in class. Similarly, you can put the onus on your students to locate cultural or even discriminatory content in textbooks and materials in order to encourage critical thinking. Berry (1997) suggests drawing on multiple resources beyond the text to present first-person interpretations, including the voices of students, because these can often balance interpretations; by relating material to personal experiences, stereotypical (or "objective") viewpoints can be challenged. She also recommends working with another instructor whose background differs from your own and who can draw upon his or her own experiences and consciousness of diversity. When using such approaches, however, Berry cautions that it is hazardous to assume "essentialism" or to appropriate others' experiences; it is also important to respect students' privacy.

Are the Assignments and Assessments Sensitive to Diversity of Learning Styles?

It is valuable to develop alternative assessments for different kinds of learners. Assessment techniques are most useful when compatible with and relevant to the backgrounds and learning styles of diverse students (Banks and Banks 1999; Lee and Carrasquillo 2006). It is particularly appropriate to diversify assessment techniques among culturally and linguistically diverse students through lab work, journal entries, portfolios, and demonstrations; such methods have been found to be highly effective with multicultural and multilingual students (Kline 1995; Allison and Rehm 2006).

Does the Course Address Affective Learning as well as Conceptual and "Factual" Knowledge?

Exercises that connect theoretical and experiential learning, for example by drawing on approaches to "active learning" (see Chapter 7 by Eric Fournier in this book), can enhance students' empathy for the perspectives of people who differ from themselves. Buckingham-Hatfield (1991) argues that participation in service and voluntary projects offers students a chance to move beyond the static gaze of traditional fieldwork practices toward an exploration and involvement with difference. Linking knowledge, emotions, action, and critical reflection are at the heart of such pedagogies. Dove (1997), for example, employed such an assignment in her classes that asked students to design and implement an urban trail, then to evaluate their designs with physically disabled students. She argues that the assignment raised student awareness about subjective perceptions of the environment, and that it subsequently improved their spatial knowledge (see Activity 10.1 on evaluating the accessibility of your campus).

Do You Include Exercises That Analyze Diversity Issues?

Some teachers have found it helpful to bring in works of fiction and other media to supplement analytical articles and to present otherwise "silent" voices. Monk (2000), for example, employed a strategy of using clips from three videos on women, work, and family in the Middle East to raise questions about how contrasting representational modes might or might not foster empathy across cultural differences. Similarly, Berry (1997) advocates employing multiple "insider" perspectives and draws upon indigenous voices in videos, audiotapes, and community narratives to implement what Miles and Crush (1993) refer to as "vent[ing] alternative voices." Brooker-Gross (1991) brought fiction into her urban social geography course by having students read a novel about a thirteen-year-old African-American girl facing racial integration of the public schools in St. Louis in 1959. Additionally, she asked students to recreate the novel's urban social geography. Rather than depending on the novel alone, the students become "detectives, using maps, demographic data, and histories and geographies of the city to locate the clues of the novel in the real world" (Brooker-Gross 1991, 40).

MAKING A CHANGE

> *Faculty and students unwittingly discuss the experiences of themselves and their peers as if everyone is [the same]. Other lives are excluded from discussions and not represented in the classroom. In so doing, structures of domination and oppression are reinforced. (England 1999, 97)*

The examples cited in this chapter illustrate various ways that instructors have transformed their teaching to be more inclusive; other examples are provided below. Drawing on their work and the wider literature on teaching for diversity and inclusiveness highlights three principles that are key to making change:

1. Plan your courses so that students recognize the centrality of diversity in all course elements: the readings and audiovisual materials, the case studies and examples you offer, the pedagogical strategies you employ, and the methods of assessing learning.
2. Find ways to make the classroom and other learning settings safe and open for all students as you design tasks, form groups, call on and respond to individuals, and set the ground rules for interaction.

3. Assess conscious and unconscious biases about people of backgrounds that differ from your own and learn how to intervene tactfully and effectively in charged classroom situations. Suggestions are provided by Sandler (2007) and University of Oregon (2007).

This chapter has emphasized that paying attention to diversity in the classroom and creating an inclusive climate facilitates the learning of *all* students. While certainly not comprehensive, the strategies I have suggested offer a starting point for you, as a new professor, to encourage an environment that supports diversity. Building diversity into one's course content and teaching practice may pose challenges, but it is essential and rewarding. Given the financial cutbacks and growing enrollment in higher education (with larger classes, higher student-to-teacher ratios, and students increasingly becoming "just a number" in the system), it is incumbent upon us as educators to consider how the streamlining of the educational system is particularly disadvantageous to traditionally marginalized students. As Villegas and Lucas put it, "teaching is an ethical activity, and teachers have an ethical obligation to help *all* students learn" (my emphasis, Villegas and Lucas 2007). It can be argued that it is only ethical for us to integrate methods into our teaching practice to support all our students, especially those who are not members of dominant majorities. The processes and critiques advocated here cannot be merely "one-offs"; rather they ought to be employed continuously, not just in a particular class or syllabus (see also James Ketchum's discussion of this issue in his chapter "Ethical Teaching in Practice").

Being self-reflective and self-critical is a lifelong project. Biases and stereotypes are not always easy to identify or admit. While the first step is recognizing the biases you do hold, the next step is to explore how they influence your attitudes and subsequently your teaching, and then to identify and implement approaches to change. I cannot reiterate enough that paying attention to diversity means always being cognizant of our own assumptions in relation to race, gender, and class, but also recognizing how students are treated differently inside and outside of class because they can be studying part-time, are in their forties or fifties, have to bring children to class, are members of a sports team, participate in reserve officer training, or are of a particular nationality. The questioning of our assumptions is something that should never end.

Of course, striving to be inclusive should not stop in the classroom. It is crucial to think about inclusion in all of our activities as academics: in the field, in our offices as we advise students (see Chapter 8 by Shelley and Proffer), as we work with staff and administrators, and when we attend conferences. Being inclusive means paying attention to diversity issues in teaching, advising, our relations with colleagues, our service work, outreach programs, and hiring practices. Look into how your own department does or does not support diversity through recruitment and hiring practices and the allocation of responsibilities and resources.

Another important strategy is to ask about the experiences of diverse faculty in your department. Nast cites the experience of one Chicana colleague who experienced consistent harassment from students on evaluations because she discussed racism in the context of Latinos. Nast writes, "she felt stranded institutionally, her colleagues being largely unequipped to recognize and deal with the racism she faced" (Nast 1999, 106; see also Pulido 2002). Similarly, Theobald's research (2007) indicates that many foreign-born faculty experience isolation when starting in a new department.

In my research on women of color and their experiences as faculty members, I found that new faculty women felt inundated with requests from students of color for extra guidance and help, which made it more difficult to find time to attend to publishing and other tasks (Mahtani 2006). They also noted that the lack of diversity can make the discipline seem unattractive to students of color (Mahtani 2004). Yet there are also efforts that can have supportive outcomes, such as the mentoring and outreach efforts of student-led groups such as Supporting Women in Geography organizations on an array of campuses, or the successful outreach programs to schools designed to enhance diversity among college students in geography and related disciplines (Rodrigue 2007). Broadening your perspectives to consider the experiences of your diverse colleagues, as well as your students, will help create truly inclusive spaces in the academy.

Additional Resources

Diversity Web, http://www.diversityweb.org, maintained by the Association of American Colleges and Universities, offers an array of resources among which are *Diversity Digest* (published online and in paper form) that addresses the intersections among various forms of difference and covers such topics as curriculum transformation, campus–community partnerships, student experiences, and faculty development and *On Campus with Women* that addresses campus climate, curricular and pedagogical matters, leadership, and other topics.

The web site of the Geography Discipline Network, http://www2.glos.ac.uk/gdn, features six guides for working with disabled students undertaking fieldwork and related activities. It includes attention to physical and mental health concerns as well as hidden disabilities such as dyslexia.

The web site of distinguished advocate and scholar of women's education, Bernice Sandler, http://www.bernicesandler.com, offers resources on such themes as freedom of speech in the classroom; how to handle disruptive classroom behavior; intervening when male students engage in negative behavior toward women; and ways for students to relate toward each other.

Many university web sites and diversity offices present information about local resources for addressing diversity. Particularly useful are the sites of

- The University of Washington's Center for Curriculum Transformation, which elaborates on perspectives presented in this chapter and offers an array of resources, http://depts.washington.edu/cidrweb/inclusive/,

- The University of Michigan Center for Research on Learning and Teaching, http://www.crlt/umich/multiteaching/multiteaching.html, which offers resources for multicultural teaching, on discussing difficult topics, and links to external sites.

A good primer on multicultural education with accompanying learning exercises can be found in the article by Moore, Madison-Colmore, and Collins (2005).

An array of articles that offer relevant approaches are identified in Monk (2000), which assesses coverage of diversity issues in over one hundred related articles published in the first twenty-three volumes of the *Journal of Geography in Higher Education*.

For those interested in addressing diversity in online courses, an article by Wang (2007) provides examples of methods to practice culturally responsive teaching through syllabus design.

References

Allison, B., and M. Rehm. 2006. Meeting the needs of culturally diverse learners in family and consumer sciences middle school classrooms. *Journal of Family and Consumer Sciences Education* 24 (1):50–63.

Banks, J. 1994. Educating for diversity: Transforming the mainstream curriculum. *Educational Leadership* 51 (8):4–9.

Banks, J. A., and C. A. Banks. 1999. *Multicultural education: Issues and perspectives*. New York: Wiley.

Berry, K. 1997. Projecting the voices of others: Issues of representation in teaching race and ethnicity. *Journal of Geography in Higher Education* 21 (2):283–89.

Brooker-Gross, S. 1991. Teaching about race, gender, class and geography through fiction. *Journal of Geography in Higher Education* 20 (3):295–304.

Brown, L. 2004. Diversity: The challenge for higher education. *Race, Ethnicity and Education* 7 (1):21–34.

Brunch, T., and A. Barty. 1998. Internationalizing British higher education: Students and institutions. In *The globalization of higher education*, ed. P. Scott, 18–31. Buckingham, U.K.: Open University Press.

Buckingham-Hatfield, S. 1991. Student–community partnerships: Advocating community enterprise projects in geography. *Journal of Geography in Higher Education* 15 (1):35–47.

Dove, J. 1997. Perceptual geography through urban trails. *Journal of Geography in Higher Education* 21 (1):79–88.

England, K. 1999. Sexing geography, teaching sexualities. *Journal of Geography in Higher Education* 23 (1):94–101.

Haigh, M. J. 2002. Internationalization of the curriculum: Designing inclusive education for a small world. *Journal of Geography in Higher Education* 26 (1):49–66.

Kline, L. W. 1995. A baker's dozen: Effective instructional strategies. In *Educating everybody's children: Diverse teaching strategies for diverse learners*, ed. R. W. Cole, 21–43. Alexandria, VA: Association for Supervision and Curriculum Development.

Kobayashi, A. 1999. 'Race' and racism in the classroom: Some thoughts on unexpected moments. *Journal of Geography* 98 (4):179–82.

Lee, D.-O. 1997. Multicultural education in geography in the USA: An introduction. *Journal of Geography in Higher Education* 21 (2):261–8.

Lee, K., and A. Carrasquillo. 2006. Korean college students in United States: Perceptions of professors and students. *College Student Journal* 40 (June):442–56.

Mahtani, M. 2004. Mapping gender and race in the academy: The experiences of women of color faculty and graduate students in Britain, the U.S. and Canada. *Journal of Geography in Higher Education* 28 (1):91–99.

———. 2006. Challenging the ivory tower: Proposing anti-racist geographies in the academy. *Gender, Place and Culture* 13 (1):21–25.

Miles, M., and J. Crush. 1993. Personal narratives as interactive texts: Collecting and interpreting migrant life-histories. *The Professional Geographer* 45 (1):84–94.

Monk, J. 2000. Looking out, looking in: The "other" in the Journal of Geography of Higher Education. *Journal of Geography in Higher Education* 24 (2):163–77.

Moore, S., O. Madison-Colmore, and W. Collins. 2005. Appreciating multicultural-ism: Exercises for teaching diversity. *Journal of African American Studies* 8 (4):63–75.

Nairn, K. 1997. Hearing from quiet students: Politics of silence and voice in geogra-phy classrooms. In *Thresholds in feminist geography: Difference, methodology, represen-tation,* eds. J. P Jones, III, H. J. Nast, and S. M. Roberts, 93–115. Lanham, MD: Rowman and Littlefield.

———. 2003. What has the geography of teaching arrangements got to do with our teaching spaces? *Gender, Place and Culture* 10 (1):67–81.

Nast, H. 1999. 'sex', 'race' and 'multiculturalism': Critical consumption and the poli-tics of course evaluations. *Journal of Geography in Higher Education* 23 (1):102–115.

Nieto, S. 1996. *Affirming diversity: The sociopolitical context of education.* White Plains, NY: Longman.

Pulido, L. 2002. Reflections on a white discipline. *The Professional Geographer* 54 (1):42–49.

Rodrigue, C. 2007. Geography diversity initiatives at California State University Long Beach: The geoscience diversity enhancement program. *Yearbook, Association of Pacific Coast Geographers* 69:160–67.

Sandler, B. R. 2007. *The chilly climate: Subtle ways in which men and women are treated dif-ferently at work and in classrooms.* http://www.bernicesandler.com/id4.htm (last accessed 20 September 2007).

Saunders, R. 1999. Teaching rap: The politics of race in the classroom. *Journal of Geography* 98 (4):185–8.

Scheyvens, R., K. Wild, and J. Overton. 2003. International students pursuing post-graduate study in geography: Impediments to their learning experiences. *Journal of Geography in Higher Education* 27 (3):309–23.

Sleeter, C. E. 1989. Doing multicultural education across the grade levels and subject areas: A case study of Wisconsin. *Teaching and Teacher Education* 5 (3):189–203.

Theobald, R. 2007. Foreign-born early-career faculty in American higher education: The case of the discipline of geography. Unpublished Ph.D. dissertation, University of Colorado at Boulder, Department of Geography.

University of Oregon, Teaching Effectiveness Program. 2007. Teaching controversial issues. http://tep.uoregon.edu/resources/diversity/methods/methodscontrover-sialissues.html (last accessed 20 September 2007).

Villegas, A. M., and T. Lucas. 2007. The culturally responsive teacher. *Educational Leadership* 64 (6):28–33.

Wang, M. 2007. Designing online courses that effectively engage learners from diverse cultural backgrounds. *British Journal of Educational Technology* 38 (2):294–311.

Research Opportunities and Responsibilities

For many academics, research is the siren song of their careers. The call of new data, an unexpected discovery, a fresh article, and the chance to make a contribution to scientific knowledge make the career journey irresistible—who wouldn't be drawn to a life like this?

Of course, most of us learn—even during graduate school—that the career journey isn't usually quite so smooth. That data set you collected? Don't be surprised if it is incomplete or inaccurate. The ethnographic study you planned for the fall semester—what if it is not approved by your Institutional Review Board (IRB) for use with human subjects until spring? Your unexpected discovery can't be duplicated in subsequent trials and the fresh article won't appear in print for two years. Suddenly, the odds of making an impact on your field seem distant.

Chances are, whether you are a graduate student studying research design or are an early career faculty member already well on your way to establishing a scholarly record, you have already experienced some of the joys and frustrations of doing research. As we have pointed out in this book, one of the things that doctoral programs do very well is prepare individuals to design, execute, and publish original research. No other aspect of academic careers is rewarded more heavily with institutional resources, funding opportunities, and career recognition.

Yet as the preceding chapters have shown, your skills as a researcher may hinge on two quite different issues. First, nuts-and-bolts issues—like how to respond to reviewer comments that seem wildly unfair; whether to publish an important piece of research in a peer-reviewed journal or in a nonjuried conference proceedings edited by a major theorist in your field; or whether to publish your dissertation as a book or as a set of articles—often are of critical importance to your career, but are topics you may have never discussed with your advisors or mentors. Second, the values you learned while in a research-focused doctoral program may be unfamiliar to the culture of the institution where you take your first academic position.

For these reasons, the chapters in this section explore some of the practical issues that affect academics doing research in all types of institutions. Though expectations for doing research will vary from institution to institution, all academics will be expected or at least encouraged to do some amount of research. With stable or shrinking budgets for higher education, a challenge for researchers at publicly funded institutions will be getting money to support data collection, research assistants, and equipment purchases. Inevitably, a tenure committee is going to ask: "What research has this candidate published? Are the journals reputable? Does the work contribute to the advancement of knowledge and theory?" And in many institutions, there has been an explosion of new collaborative models of funding and doing research that cuts across traditional disciplinary boundaries, with real consequences for how graduate students and faculty are prepared and evaluated during the early stages of professional development.

OVERVIEW OF THE CHAPTERS

In Chapter 11, "Preparing Competitive Research Grants Proposals", Patricia Solís argues that effective proposal writing is about more than being a good writer (although writing strategies, as Patricia notes, are certainly essential for crafting a successful proposal). Effective proposal writers are also armed with clear ideas and an awareness of how their research interests complement the goals of potential sponsors. In particular, Patricia explains how a competitive proposal is one that, fundamentally, achieves coherence: a problem statement that is clearly rooted in related literature and connected to a set of research questions, a description of data sources and analytical methods appropriate to the questions and data, and a dissemination strategy that will help the research achieve broader impacts in the discipline and perhaps beyond. The activities paired with this chapter can help you formulate an effective proposal and ensure all of its myriad parts, when considered together, coalesce into a convincing plan that can be delivered on time and on budget.

Regardless of whether you need external funding or can rely on other sources, your employer's IRB (or equivalent) will require a review of human subjects protections prior to permitting a research project to be implemented. But as Iain Hay and Mark Israel show in Chapter 12, the IRB process is only one aspect of research ethics that academics need to know about. In fact, human subject protocols can seem quite straightforward compared with the unanticipated situations that oftentimes surprise researchers in the social and environmental sciences. One of Iain and Mark's activities features several case studies aimed at promoting awareness and reflective dialogue on issues ranging from research authorship and confidentiality to questions about the disclosure of researcher identity and conflicts of interest.

One of the pleasures of successfully completing a research project is writing up the results for publication. Stan Brunn's chapter on academic publishing touches on many of the mechanical dimensions of writing manuscripts, but delves much further into the often mysterious, behind-the-scenes activity that occurs after manuscripts are submitted: the criteria that editors use to evaluate the quality of submissions, the role and essential service of peer review, and the challenges of revising manuscripts in a manner that is responsive to the concerns of your reviewers. Understanding this process is an essential step toward getting your work published, and the chapter's activities are designed to demystify the sometimes intimidating procedures and frustrating expectations of academic writing and publishing.

At some point in your academic career, you may feel tugged toward literature in disciplines that overlap your own research interests. Whether for practical necessity or mere intellectual curiosity, as someone who studies fluvial geomorphology you may find yourself browsing the academic journals in geology just as often as journals in physical geography or hydrology. Or if you study transportation systems, you may rub shoulders with civil engineers and urban planners at that next GIScience meeting. A conference on climate change may draw as many palynologists as meteorologists. Such are the opportunities outlined by Craig ZumBrunnen and So-Min Cheong in Chapter 14 on "Working Across Disciplinary Boundaries." As one of the defining trends in the modern academy, interdisciplinary research has greatly expanded opportunities for professors to engage in collaborative work with colleagues outside of their home department and institution. But effective interdisciplinary work requires habits of mind to which many new professors are unaccustomed: collaboration and group work, sharing data and writing responsibilities, and getting to know the cultures of new disciplines are some of the ways that interdisciplinary research casts the topics in this book in a whole new light. The experiential team-based activities accompanying Craig and So-Min's chapter are intended to assist you in learning some of these habits.

As you read the following chapters you will also see how the themes hearken back to many of the ideas expressed earlier in this book. Indeed, you can mobilize the skills emphasized in this third section on research opportunities and responsibilities to improve your overall effectiveness as a teacher, advisor, and colleague. Many of your own students will be interested in writing a journal article, submitting a grant proposal, or taking courses offered by other departments. Establishing a balanced pattern of research and writing will open up pockets of time that you can use to learn about a new educational technology or volunteer to serve on a departmental committee. The principle of coherence is central to both effective proposal writing and aligning course objectives with assessment strategies. Developing ethical practice in your research, as in your teaching, will sensitize you to the larger context of the academic enterprise and the evolving roles and responsibilities of academic professionals.

The chapters in this third section, then, provide more examples of why, by virtue of the interlocking nature of academic work, our academic culture must change to view the preparation of future faculty as foundational to the health of academic departments, institutions, and disciplines.

CHAPTER 11

Preparing Competitive Research Grant Proposals

Patricia Solís

Beyond the academic duties of teaching, research, publishing, and service, expectations that faculty should generate external funding are increasing—and are increasingly important. Tenure and promotion evaluations of faculty may or may not explicitly review dollars won, but are clearly influenced by successful grant-getting as a measure of contribution to one's field, especially for peer-reviewed proposals. In nearly every type of academic institution, from small liberal arts colleges to state public universities, the pressures of shrinking state budget allocations coupled with increasing proportions of organizational revenue coming from outside sources (Lee and Clery 2004) make grantsmanship an ever-critical aspect of professional life for tenure-track research faculty, lecturers, and part-time faculty alike. Graduate students, as well, seek outside monies to conduct their research activities, not only to garner additional financial support, but also to demonstrate their abilities in fund-raising and gain early grants experience that will serve them well in their budding careers.

Even if a grant is not awarded, the process of writing research grant proposals can help you solidify your research ideas and make them tangible, realistic, and programmatic. Proposals, funded or unfunded, can themselves become sources for publishing journal articles (a point reinforced in Chapter 13). Writing to external sponsors can aid you in thinking through your rationale or finding ways to connect with others in the scientific community or with the broader society—whether you end up with the money or not.

This chapter aims to give you some advice for how to do that successfully. Admittedly, there have been a great many articles and books written on the subject of grant proposal preparation, and at the end of this chapter you will find an annotated list of those which I deem among the best. These resources adeptly cover issues such as choosing research topics, mechanics of writing, necessary elements of proposals, hallmarks of successful proposals, obstacles to overcome, strategies for locating funding sources, perspectives of sponsor agencies, and effective research design. In this chapter I include a focus not only on practical recommendations that will help you prepare competitive research grant proposals but also on habits of mind and action that will prepare *you* as a researcher who is successful at generating external funding.

Ten actions you should take are listed below. This list, of course, does not represent everything you need to know but will help you develop your own strategies and tactics for success to incorporate into your professional practice:

1. Start with a good problem.
2. Create the right fit.
3. Assemble a winning team.
4. Design to deliver.
5. Be perfectly persuasive.
6. Make it real.
7. Demonstrate your unique value.
8. Go the extra mile.
9. Achieve and communicate coherence.
10. Live it.

This chapter explains each of these ten imperatives and includes some suggested concrete actions you can implement either as a novice to begin preparing winning proposals or perhaps as a seasoned grant writer trying out some new ideas and tools to refine your skills—ultimately to become a more effective research grant proposal writer. In box diagrams throughout this chapter, I highlight proposal excerpts contributed by graduate students and faculty whose proposals were funded by the National Science Foundation [NSF] (these proposals appear in chapter appendices on this book's web site).

Most research grant writing skills must really be learned by practice, and unfortunately, trial and error. To this end, in-depth activity guides are provided for two of the actions, "Start with a Good Problem" and "Achieve and Communicate Coherence." The first is designed to help you turn your research idea into a clearly communicated thesis paragraph. The second activity is designed to aid you in the revision of a relatively complete proposal draft, providing a set of relational questions in the form of a matrix of elements that your proposal should contain.

TEN ACTIONS YOU SHOULD TAKE

1. Start with a Good Problem

No amount of writing skill will compensate for a poorly developed research problem. A good problem is more than just a topic; it encapsulates some kind of tension, contradiction, unresolved issue, challenge, question, or even a mystery of sorts. A particular problem makes for a good proposal if it is very specific, precise, and focused. You should be able to clearly express what this research problem is in no more than three to five sentences.

Good problems lead to good questions, which underlie the scientific enterprise, be it theoretical or applied, quantitative or qualitative. The objective of a research project is always to answer those questions. This sequence of problem–question–purpose should be as explicit as your thesis. Clarity, lucidness, and a sharp thesis will allow you to develop a sound project and a convincing proposal. It will leave your reviewers with something to remember after reading all of the many other competing proposals.

A good problem is properly contextualized. A critical element of grant proposals is a summary of the current state of knowledge that is supported by references to relevant scholarly publications. You should carefully and explicitly position your work relative to this context. This entails a different approach to writing than would a traditional literature review, especially as you consider that most likely not all of your reviewers will be familiar with your specialized literature. A good rule of thumb is to precisely answer this series of questions: "What is already known? What don't we know that is worth finding out? What will be learned from this research?" Or "What [have others] done to address the problem and why wasn't that sufficient?" (Gerin 2006, 73).

To communicate this contextualized problem, it is helpful to consider three general models of what research is supposed to accomplish with respect to existing knowledge. These include that research might (1) advance an existing line of research, (2) resolve a contradiction (Box 11.1), or (3) develop a new line of inquiry (Box 11.2).

There are certainly other functions of research, but these three are common models to begin thinking about how your project contributes to the academic enterprise of knowledge production. Activity 11.1 and Appendix i provide additional discussion and opportunities for you to explore these ideas.

BOX 11.1

The following excerpt of a funded proposal from Lawson & Jarosz (Appendix v) eloquently demonstrates how research can start as grounded in an important empirical problem and can contribute to our understanding by resolving a contradiction:

> "All across the American West, rural families are dealing with rapid changes to their livelihoods and communities. Many rural places are experiencing dramatic social and economic transformations as urban middle and upper class migrants bring new politics, new aesthetics and new economic activities into contact with longer-term resident populations (Nelson, 1999; Beyers and Nelson, 1999; Rudzitis, 1997). These changes have brought the *paradox of both income growth and rising poverty rates.* Along with these new income distributions within rural counties, shifts in the class composition of communities and new social tensions between residents are emerging. While there is a growing volume of research on the demographic and economic dimensions of these changes (Rasker and Alexander, 1997; Nelson and Beyers, 1998; Shumway and Davis, 1996; Barrett and Power, 2001), *less attention has been devoted to the social and cultural tensions surrounding this process* of rural restructuring and how poverty is understood within this context of transformation and change (Rudzitis, 1993; Cloke, 1997; Nelson, 1999)." *(emphasis added)*

Note that the use of terms like "paradox" and "tensions" emphasize the function that this research intends to perform and help create a focus on the interplay of factors that the study will address.

BOX 11.2

This proposal exemplar simply and elegantly opens up space for a new line of inquiry (Patel, Appendix vi):

> "In a recent review, Tuan (2004) contends that "cultural geography remains almost wholly daylight geography" (Tuan 2004, 730) and that more attention needs to be given to the "after hours." This contention makes particular sense as the "second shift," namely a night shift labor force, emerges in the global economy. The hypergrowth of the transnational call center industry in India is the quintessential example of this nightscape . . ."

2. Create the Right Fit

Like people, institutions have their own personalities, and funding institutions do as well. Whether they are private, federal, state, foundation, or other kinds of organizations, each has its own specific mission, goals, and program objectives. Each sponsor organization also has particular preferences on operational protocol, including whether it invites unsolicited proposals or not, whether it issues calls for proposals that are clearly or vaguely defined, whether it prefers formal or informal inquiries, whether program officers should be contacted prior to deadlines, and other tacit protocol. You should make it an integral part of your research proposal preparation process to find out about the individual characteristics of potential resource providers. In other words, do your homework before you write. Many organizations now have easily accessible information online where you can research past funding history and current priorities. If information about previously funded projects is available, study it. If only the contact information of their previous awardees is listed, consider talking to funded principal investigators (PIs) about their projects and experiences with the sponsor organization. Do talk to program officers if possible, but have a clear objective for your conversation.

There are also general guides available about how the funding process works with sponsors from different sectors. For instance, it is very important for aspiring academics to know about the federal government in general (see CFDA n.d.), given that the federal government remains the largest research sponsor to U.S. universities (Lee and Clery 2004); and about the National Science Foundation in particular (see NSF 2005). Private enterprise donors are increasing in importance for researchers, and resource requests to businesses call for a specialized approach that understands the sector's perspective (see Schumacher 1992). Similarly, foundations often require more insider knowledge, and they conduct calls that are unlike the competitive proposal processes of public institutions (Geever 2001).

The purpose of knowing your potential funding organization well is to be able to write to the right audience and create what I call the "Right Fit"— finding the overlap between your own research agenda and the goals and mission of the funding organization. Alternatively, you might think of a good fit as a temporary alignment in the same general direction to advance goals together. You will need to consider how to package your research idea, which as Michael Watts (2001) notes, is not the same as to "compromise" it.

The way you go about searching for potential funding opportunities is itself an opportunity to ensure the right fit. One approach is to follow the path of leaders in your field. The Curriculum Vitae (CVs) of experienced researchers in your area are a gold mine of potential funding sources for your own research. Many scholars now publish their resumes online or at least some kind of list of recent projects. Look up the name of a senior scholar in your subfield, using a search engine to find their home page, or find them through their department's faculty page if you know their institution. Locate their CV or resume, and make a list of all of the names of organizations where their research has been funded or where they have conducted projects in collaboration. Odds are that your research will be eligible for funding from those same organizations. They are also more likely to be a better fit than less-focused searches.

Organizing and systematically prioritizing which funding opportunities align best with your work—rather than jumping at the nearest deadline—will help you focus on the best fit and, by extension, the highest probability of success. Create a spreadsheet of possibilities from a variety of sources, listing the potential sponsor organizations' name, address, web site, contact person, e-mail, telephone, published funding program priorities, grant funding ranges (minimum to maximum amount of awards), eligibility criteria, deadlines, and so on. Annotate each record with a reference to a particular research problem or question (not just a topic) that might be developed into a proposal. Rate the likelihood you believe you have in receiving funds from each potential source, using a percentage scale. This is a bit of an instinctive process, weighing a multiplicity of factors such as the level of competitiveness, how good the "fit" is, how amenable the timeframe is to your research agenda, how well you match the eligibility profile, whether the request limits will meet your resource needs, and other factors. In this way, you can prioritize the list of possibilities according to the likelihood of success and your own preferences for research questions.

For discovering and creating the right fit as you design and write the proposal, you might find common ground in one or more of the following areas:

Organizational mission: There may be some shared vision between the goals of the organization and your respective research goals. These may not align perfectly, so the objective is to discover where they overlap and exploit those connections fully (Box 11.3).

Targeted beneficiaries and research subjects: Pay special attention to shared constituencies and their needs (*e.g.*, both you and the funder are interested in youth populations) or relationships between respective constituencies (*e.g.*, your interest in youth and their interest in retired professionals possibly connecting through mentoring relationships).

Operational approach: The means of executing your projects may coincide with those that sponsors prefer or promote (*e.g.*, the use of geographic technologies, mapping, communication technologies, community participation, or shared markets).

Mutually beneficial deliverables: What might your project produce that serves both you and the funding organization? (*e.g.*, a tourism brochure for the organization that includes a community map you seek to create).

Global objectives: What global or international targets might you both care about? (*e.g.*, investigating progress toward the U.N.'s Millennium Development Goals).

BOX 11.3

This example proposal (Solís, Appendix vii) generates and communicates a rationale for the project to be conducted within a professional society, deviating from the norm that the agency makes awards under this program to university-based researchers. It first establishes common ground in the missions of the organizations:

> "As producers of knowledge in a globalized society, scientists and scholars must reckon with a world where economies are increasingly connected, where country boundaries are ever-shifting, where communication technology enables information to traverse the globe with great speed, where workplaces are increasingly internationally distributed, and where cities and towns are growing in ethnic and racial diversity. These phenomena impact the activities of knowledge producing enterprises, a fact which has prompted research-oriented organizations such as universities, R&D business, and federal agencies to promote international research collaboration (IRC) as a means to build intellectual capacity and increase competitiveness, among other goals (See National Science Foundation NSB 03-69; NSB 00-217; American Association for the Advancement of Science/Teich 2000; Social Science Research Council 2000)."

Next, it explicitly evokes a shared constituency:

> "Because the disciplines are the intellectual home of faculty members, disciplinary professional associations can play a central role in any effort to understand and enhance the international dimensions of academic research (Lawson 2005). However, research on IRC has focused on primarily two sets of actors, the individual scholar and universities or groups of institutions and seems to neglect the influence that disciplinary professional societies wield or potentially could bear upon facilitating productive IRC."

And finally offers a mutually beneficial deliverable in the form of:

> "An International Model that can be applied to other disciplines and other regions to advance scientific cooperation in research and scholarly inquiry, to be broadly disseminated . . ."

Communicating in the language of the sponsor is a critical part of creating the right fit for your proposal. Scholars should recognize that the style of writing for a grant proposal differs sharply from that of an academic journal article, for instance, and different modes of expression are called for. Even proposals to scientific funding agencies such as the NSF should avoid scientific jargon and use an appropriate style that reflects the manner in which the funding agency communicates. When key terms are used in the proposal announcement, integrate those same terms in your proposal. Don't leave reviewers guessing if your research meets their specific objectives; tell them exactly, in *their* own words how it does. To practice, read through an entire proposal announcement carefully and identify "buzzwords" that in the language of the sponsor express important key concepts significant to their organizational mission or the goals of the granting program. These might be words, word combinations, or phrases. List them, and then define each of them using the kinds of language that your research would normally use to express ideas to create a kind of glossary, a handy reference when you are writing and revising your proposal.

Beyond the horizon of one particular grant proposal, long-term success in generating resources for your research rests upon your ability to build and maintain mutually beneficial relationships with sponsors. You should be willing to invest in developing these relationships over the long term. Start planting seeds with a good understanding of who the sponsor is and with what language they communicate. Resource requests, including everything from informal inquiries to formal proposals, themselves often both draw from and generate relationships, whether they ultimately are short or long term, whether intimate or formal or casual. Doing it right may make the difference not only for a positive response to your current proposal but also for the long-term success of your research agenda.

3. Assemble a Winning Team

The saying goes that "no man is an island." Grant preparation is no different. Assemble a winning team on your side to increase your chances of success. This advice holds for experienced faculty, early career faculty, and graduate students alike.

There are at least three kinds of help you might need in preparing your research grant proposals. Firstly, you should early on ask for help from your university's Sponsored Research Office or Program (SRO), where they can provide a wealth (and often literally a library) of information, including granting agency directories, model proposals, writing services, human subjects advice, and so on. In most cases, consulting with the SRO from the start can also help you get through university review more easily and quickly.

University review is often required because most grants are contracts between the granting agency and the university (or a unit within the university), not between the agency and the researcher. A typical university review will consist of a series of forms and procedures, including some kind of researcher information form, a proposal or abstract form, budgeting forms (including your final budget with justification), human subjects' approval, and other information. Generally a litany of signatures is required, from you as the PI, your department chair, your dean, the university office of grants and contracts, and finally the SRO. If your university does not specify the time frame by which you must submit all materials prior to the granting agency deadline, it is recommended that you provide your complete written proposal (and all related documents) at least three to five business days before, but at some universities a much longer approval period is needed. You should check with your office well ahead of time to know what their internal deadlines and requirements are. Any research that involves human subjects must also undergo a review by your university's Institutional Review Board (IRB), for which you should leave at least a couple of weeks if the granting organization requires final approval prior to submission (see Chapter 12 by Iain Hay and Mark Israel in this book for a detailed treatment of IRB issues). Finally, a common required element of the research grant narrative is some general information about the applicant institution (i.e., your university). Your SRO should be able to assist you with standard responses to those requirements.

Secondly, you might seek out the aid of a seasoned grant writer who can help you keep on task and on time and cover all of the bases. This person can play the role of practice reviewer for later drafts of your proposal. If willing and available, scholars experienced in proposal writing can serve as critical mentors as you develop your skills.

Thirdly, when appropriate, you might invite these same scholars to team up with you on the research project as a co-principal investigator. Nothing breeds success like success, and if you can develop collaborations and partnerships with other successful researchers, you will see your own success expand. By building a team, you may also become eligible to compete for larger funding programs.

If you are not quite ready yet to call a team to arms around a particular research proposal, you could simply make an appointment with a scholar or two in your subfield who are experienced in writing and getting grants. Interview them, share your ideas and questions with them, invite them to mentor you on specific aspects of the process, or generally consider developing joint proposals with them. If you do not find an experienced colleague, you could also benefit greatly from doing the same even with a less-experienced peer.

4. Design to Deliver

The actual research activities and methodologies that you will undertake should be very carefully designed and clearly organized to instill confidence in your reviewer that you can deliver on your promises. As Krathwohl (1988, 15) advises, "while projects typically start with an idea, sponsors fund activities, not ideas." Focus on what you will *do*. Keep your scope realistic and doable within the granted time period. Think carefully through your entire plan, step-by-step from start to finish to ensure that you will be able to administer the research project on time and on budget. Practice good time management (see Chapter 1 by Ken Foote), not only in your writing process but also in the design of the project; determine how long and at what point in time each of the set of activities will need to occur.

To be clear about what you will do when, how, and with what resources, you should as a matter of practice always include a separate section called "project design" where you identify the specific *activities* needed to carry out

BOX 11.4

This example timeline for a multi-year project is clear and succinct, and summarizes the more detailed methodology section (Song, Appendix viii). Notice how each quarter's activity domain is focused and follows from one time period to the next:

Year one	1st Quarter	Data acquisition and preprocessing, including geometric correction and radiometric calibration, to build a GIS database.
	2nd Quarter	Finish preprocessing and GIS database development, test running RHESSys for the Blackwood Division of Duke Forest.
	3rd Quarter	Start developing spatial algorithm to map tree size and density with high-resolution remotely sensed data.
	4th Quarter	Validation of tree size and density derived from 1993 B&W DOQQ, 1998 DOQQ and recent Ikonos/QuickBird data.
Year two	1st Quarter	Modify RHESSys to take tree size and density over the landscape to simulate carbon cycle.
	2nd Quarter	Calibrate RHESSys with flux tower measurement and error analysis. Start transferring GORT-ZELIG to the local area.
	3rd Quarter	Continue work on model simulation with GORT-ZELIG and develop algorithm to map stand ages with ARTMAP.
	4th Quarter	Validating stand age map and integrating stand age with RegCarb to simulate carbon cycle.
Year three	1st Quarter	Error analysis for age effect on carbon cycle; mapping subpixel tree cover and LAI with MODIS/MISR.
	2nd Quarter	Evaluate subpixel LAI and tree cover products; test running a light-use-efficiency model.
	3rd Quarter	Error analysis for subpixel LAI and tree cover with Radiosonde measurements and other simulations.
	4th Quarter	Publication preparation, and project report.

your research methodology, a *timeline* (see Box 11.4 and Appendix ii), and a bulleted list of *deliverables* or products or outcomes expected—even if the call for proposals does not specifically request these items.

Designing to deliver also means that you will be able to realize your project with the requested resources. Above and beyond serving as your formal request for a particular dollar amount, your budget section should demonstrate your resource plan. The single most important key to designing a good budget you can live with and get approved is to be very clear first about what you will *do*, then ask for the resources you will need according to each activity. Ask yourself for each activity, what do I need to perform this task? Convert how much time it will take into dollars, since most budgets also pay for salary or wages.

You may ultimately have to negotiate a final budget with some sponsors, so be sure that each item in the budget is fully justified and necessary, or you may find that it gets cut. Most novice grant writers need some guidance on deciding how much to ask for and how to ask for it. A good rule of thumb is to use the guidelines and limits in the call for proposals, remembering that you do not have to ask for the full amount available, but you should ask for the amount you need and can appropriately justify. For travel expenses, look up current federal government guidelines for per diem rates; for equipment expenses, get some quotes from vendors to provide evidence of costs; for indirect rates that differ widely among universities, be sure to check with your SRO for proper procedure and documentation.

In short, basic criteria for designing a budget that delivers include the following:

- Is each item eligible for funding according to donor rules?
- Is each item necessary for the project and linked to a particular activity?
- Is each amount properly justified?
- How much "bang for the buck" can your project deliver? Put your request in perspective according to what impact your research will have.

For more details on budgeting basics and how to test a budget to ensure that it is reasonable and sound, see Henson (2003) and The Foundation Center (2006).

You should strive to meet the expectations of the granting organizations, or at least your reviewers with what you ultimately propose to deliver. Expectations, of course, differ by field, with research in the humanities, social science, and physical science having their own unwritten codes of expectations of their communities of scholars. Conform your proposal to the appropriate context. If you don't know much about that set of unspoken assumptions, talk to a mentor. In any case, this usually means that you should deliver some kind of impact—be it impact on practitioners of the field, impact on our knowledge about the problem, or impact on a community affected by the issue. Clearly specify what that impact is and show how you will achieve it.

5. Be Perfectly Persuasive

Grant writing uses persuasive communication, aiming to convince the reader that the project is worth investing in. This means that you should use carefully crafted arguments backed up by evidence, not unfounded assertions that your reviewers might doubt or worse, challenge. It also means that the language should exude confidence, eschewing grammatical formulations that are overly reliant on "might" or "could," opting instead for simple present or future tenses as often as appropriate, like "is" and "will be." To convey the sense of action that you will perform as you carry out your project, be sure to avoid the passive voice unless absolutely necessary.

Excellent persuasive proposals distinguish themselves by anticipating opposing arguments or identifying points of contention in the research plan. By preparing a positive response to defend likely criticisms, rather than avoiding them, you demonstrate thoroughness in your thought process and can convince reviewers that might otherwise remain skeptical about how your research will contribute to the field (Box 11.5). Mentors and colleagues can help identify opposing arguments. Or, consider reviewing your proposed work in a seminar where the discussion might raise ideas about possible responses to your work.

Being persuasive is not only about what you say but also about how you say it. Be sure that your document is perfect, error free, and looks great. First and foremost, conform *exactly* to the formatting specifications in the call for proposals. Use the very same headings and subheadings for proposal narrative elements in the same order as requested (*e.g.,* Introduction, Significance, and Methodology). Don't cheat on line spacing or margins; use standard, black color, and readable fonts and font sizes (never less than 10 point) and stick with the same choice throughout all of the proposal documents. Leave plenty of white space and break up long paragraphs of text. Use but don't overuse bullets, tables, and graphics to draw attention to particular elements of your proposal. Don't rely on automatic spell check in word-processing programs to catch all spelling or grammar errors. Don't allow widow or orphan lines, always number your pages, and use a common paragraph justification for the whole document.

When you finish, step back from your document to check for possible formatting problems by using the "zoom" feature of your word-processing program—scroll through to look for inconsistencies (Box 11.6). Print out a hard copy to see what it looks like, even if you will be submitting electronically. Leave enough time in your writing schedule to let your proposal gather a little dust and reread it fresh again, or ask a colleague to proofread for you.

In short, a good rule of thumb is to avoid any kind of presentation problems that might irritate your reviewer or distract from your content. Similarly, follow all submission guidelines precisely and on time. Leave extra time for using online submission systems, especially if it is your first time preparing an electronic proposal for that agency. By paying attention to these details, you not only produce a polished proposal, but you will also instill confidence in your reviewer that you can deliver a well-conducted research project.

BOX 11.5

This excerpt aptly addresses a potential concern from reviewers (Song, Appendix viii). After introducing the concept of "successional stages" on which the research methodology relies, the author anticipates and dismisses a possible objection to its use:

> "Forest succession is closely related to tree size and density, but it also incorporates the individual replacement process. Stands at different forest successional stages are usually composed of trees at different ages, and sometimes different species composition. Though the mechanisms *are debatable* (Yoder et al., 1994; Ryan and Yoder 1997), *the fact* that forest productivity strongly depends on its successional stage *is well accepted as it is confirmed* in numerous studies from different perspectives (Birdsey et al., 1993; Law et al., 1999; Barford et al., 2001; Law et al., 2000). Therefore, knowledge of forest successional stages over space should lead to improved estimation of carbon budget over the same area." *(emphasis added)*

Similarly, this example silences a possible contention over the choice for her study area by addressing it in advance (Patel, Appendix vi):

> "Arguably, as call center operations emerge throughout India, the question becomes why focus on Mumbai versus Bangalore, Chennai, or Hyderabad. When asked 'Why setup in Mumbai versus Bangalore?' Sharon, a call center executive, contends that Bangalore is the IT hub, but not necessarily the call center hub. The presence of an educated, English-speaking population and the space available to build call centers in the outlying areas of Mumbai are the key magnets drawing companies to this area. Mumbai is also viewed as more cosmopolitan and professional, and is ahead of Delhi in terms of fiber-optic connectivity and its electricity infrastructure (Patel 2002). At the same time, during pre-dissertation research I discovered that some families are hesitant about women working for a call center in Mumbai because, unlike Bangalore, Mumbai is viewed as a city of ill-repute, danger, and sin. In this context, focusing on Mumbai provides a complementary understanding of how the local conception of a cityscape intersects with the global demand for a night shift labor force."

6. Make It Real

Related to the idea of persuasion is the strategy of making your project real. The discipline of geography, for example, has a distinct intrinsic advantage of being intimately connected to the actual phenomena that we study, phenomena that generally intersect with our everyday lives in a real way, whether

BOX 11.6

These sample pages demonstrate how to "zoom out" and check that your work looks professional. The text to the right is too dense, and is not justified cleanly. There is not enough white space to smoothly lead your eye through the narrative. Ten or fifteen pages of this style will surely irritate your reviewer.

Judicious use of graphics, bullets, boldfacing, double justification, and white space greatly enhances the readability and flow of your narrative, as these two sample pages show:

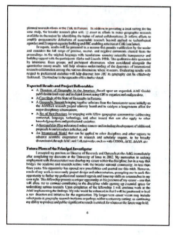

they be about income distribution patterns in a major city or linkages between rural land use and groundwater quality or statistical evaluations of global climate change models (see Box 11.7). Even theoretical research often draws from and has implications for very real everyday matters. Take advantage of this perspective to connect to your reviewers by providing them with real-life examples and connections that avoid jargon. Infuse your narrative with *one or two* (don't over do it!) carefully selected maps or photographs that validate the adage of a picture being worth a thousand words. Writing with passion (but not hype) also makes it real for your reviewers, as Abler (1989) so elegantly enjoins. Flip through newspapers and magazines for inspiration on compelling examples. Explain your unique geographic work in a way that will implant an image into the minds of your reviewers to remember as they wade through the other scores of proposals they still have to finish reviewing this afternoon.

BOX 11.7

"Many decades ago, Americans observed the effects of wildfires on the composition and structure of hardwood forest environments (the kind of natural environment where most people lived at the time). Based on those observations, people drew the plausible conclusion that fires in nature were harmful, and that policies were needed to "protect" forests from wildfire. Well-intentioned applications of fire suppression policies in other environments – e.g., the Plains grasslands, Rocky Mountain coniferous forests, or southern California chaparral – have proven unwise.

We now know that fire is an integral part of many natural environments, and that the suppression of normal fires in those environments will lead inevitably to an altered ecosystem, often accompanied by a buildup of fuel and a more catastrophic fire. Similar examples could be cited in such diverse fields as urban planning, natural hazards, counterterrorism, agricultural biotechnology, infectious disease control, and energy policy. In each case, activities and policies that work well in one place may be less effective or even counterproductive in another place."

Excerpted from Solem 2004

7. Demonstrate Your Unique Value

The proposals that are successful are the ones that stand out from the crowd. Explicitly state what is special about your work. Usually every call for proposals requires a special section to describe the "significance" of the project, where you identify what is unique about the proposal. Brainstorming answers to the following questions may help you pinpoint what aspect or combinations of characteristics are valuable in the work you plan to do

- What is special about the problem I have chosen to address?
- Is my work conceptually innovative?
- Does the field know less about my case than other similar cases elsewhere?
- Have there been any recent critical breakthroughs in the field that my work draws from or seeks to build upon?
- What about my methodology might represent a contribution to the field?
- How is my approach different from the way others have looked at the issue?
- What is the payoff? (Przeworski and Salomon 1995)
- What is new about what my expected research results?

Use words that signal difference, uniqueness, and the special quality of your work (Box 11.8).

BOX 11.8

This example excerpt (Song, Appendix viii) compares the proposed approach to others to demonstrate its unique value:

"We propose to expand the scope and depth of existing studies at Duke Forest by scaling up carbon fluxes from stand to landscape through integration of remote sensing, ecological modeling and ground observations. Our scaling up strategy *differs from the traditional* 'big-leaf' model as we explicitly incorporate spatial vegetation heterogeneity into ecological models to simulate landscape carbon cycle." *(emphasis added)*

Another example integrates compelling vocabulary to show its special character (Wasklewicz, Appendix x):

"The proposed education and research activities are *a significant shift away* from the compartmentalization of techniques and concepts found in many disciplines. A holistic approach, like the one proposed, will compel students to develop sound habits in project design, fieldwork, data collection and management, analysis, synthesis, and articulation (written, spoken, and visual). A student immersed in this learning environment can provide *innovative approaches* to broaching integrative subject material. The application of these systematic practices will permit students *to go beyond* lecture and lab to conduct publishable original field- and computer-based research because they have a clearer view of science and its applications" *(emphasis added)*

Certainly, as Przeworski and Salomon (1995) point out, "disciplinary norms and personal tastes in justifying research activities differ greatly: Some scholars are swayed by the statement that it has not been studied (*e.g.,* an historian may argue that no book has been written about a particular event, and therefore one is needed), while other scholars sometimes reflect that there may be a good reason why not." In any case, justifications should be based upon evidence and argumentation rather than assertions or opinions about what is appealing about the subject matter.

Generally, another effective approach is to capitalize upon trends in your field or recent events worthy of note (Box 11.9). For instance, in response to interdisciplinary calls for proposals, the broader trend of a "geography rediscovered" can give contextual importance to your work. Over the last decade, the discipline of geography has undergone a renaissance that has moved it to the academic center (NRC 1997; Pfirman and the AC-ERE 2003) and has generated an importance for geographic research in society at large (Richardson and Solís 2004; NRC 2006). Consider the fact that geography majors have increased by 32 percent in the past 5 years (NCES, various years), dozens of new geography programs have been initiated, including at

---

BOX 11.9

===

Wasklewicz (Appendix x) positions his research as a timely intervention, taking advantage of new developments coupled with a rich empirical tradition:

> "Form analysis in geomorphology has *languished for decades*, mired in studies of 2D shape as opposed to interpreting 3D and 4D landform characteristics. This produced a situation whereby our current understanding of continuous terrain is not equated with repeatable, measurable form attributes, but rather with qualitative observations or simplified empirical interpretations. *Recent developments* in geomorphometry have promoted the concept of numerically characterizing form by analyzing geomorphometric structures. Structure arises from a quantitative understanding of the spatial arrangement of morphometric point data and represents a numerical signature of the topographic form. *Historical analyses* of alluvial fans, which have produced *a solid literary foundation*, the ubiquitous nature of fans, and an exposed surface expression of fans *make them an ideal feature for establishing* a morphometric structure approach." *(emphasis added)*

---

prestigious universities such as Harvard and Howard (*AAG Guide*, various years), geographic information technologies are diffusing rapidly in all sectors of the economy (Gewin 2004), and job opportunities are multiplying rapidly as demand rises for workers who are globally literate, knowledgeable of geographical concepts, and skilled in interdisciplinary research methods (Jackson 2005). As you make your case for the importance of your own research, it may serve your purposes to point out one or more of these aspects of a growing discipline (Pandit 2004). A few well-placed bullet points can impact your reviewers and increase the likelihood that they will understand your research as significant.

If you choose a problem that is unique because it is a topic of current salience, you should take care to convince your reviewers that "such topics are not merely timely, but that their current urgency provides a window into some more abiding problem" (Przeworski and Salomon 1995) (Box 11.10).

However, beware of the fact that "hot topics," whether theoretical or applied, will likely also attract more competitors. If everyone is writing about it, you may be better advised to develop or stick to a quality research niche that is your own. "By the time you write your proposal, obtain funding, do the research, and write it up, you might wish you were working on something else. So if your instinct leads you to a problem far from the course that the pack is running, follow it, not the pack: nothing is more valuable than a really fresh beginning" (Przeworski and Salomon 1995). Doing so will not only distinguish your particular proposal but will also set you apart as a researcher in your own right.

BOX 11.10

Lam, Campanella, and Pace (Appendix iv) explain the intellectual merit and broader impacts of their proposed project in the immediate aftermath of Hurricane Katrina. Note that both short and longer term benefits of their work are featured in this justification:

> "Very little research has focused on collecting *time-critical, empirical data* on how businesses make spatial decisions on whether they remain or relocate after a catastrophe, *especially a catastrophe as deep and wide as we have seen* that affects an entire metropolis of New Orleans. The time-critical data that we collect for this project will provide *unique information* on how decisions among businesses are made in this *unprecedented* case. The coupling and tracking of street and telephone surveys over time will provide vital information for research on human-social-economic dynamics over space and time. The data we collect will also serve as an important *benchmark dataset for subsequent research* and for comparisons with other studies (e.g. studies on decisions made by individuals) . . . The time-critical, integrated GIS data set collected in this project can be made available to other researchers and planners, and can be used as a *basis for further related research*, such as modeling the impacts of Katrina on health, poverty, and crime. Our preliminary analyses of the data will be published and widely disseminated. Our data will provide a first-hand, *rarely captured view* of how a city recovers, literally from ground zero, and how businesses make decisions in post-catastrophe uncertainty. This information will help governmental and planning agencies in devising effective policies for economic recovery in the region." *(emphasis added)*

---

8. Go the Extra Mile

Increase your competitiveness by including one or more strategic elements that may not be required but can help round out your proposal. While some of these ideas apply better to faculty proposals, including others in student proposals may demonstrate early professional maturity. All should be applied to enhance your work, not distract from it. Be sure to check that your target sponsor agency allows these items in proposals; some sponsors have very strict guidelines. At best, incorporating these strategies regularly can help you transform individual research projects into a solid research program.

An *advisory board* is a group of people who can help you ensure quality and provide a sounding board for ideas and problems. If you can identify and convince experienced scholars to agree to serve on such a board (pending acceptance of your proposal), referring to their tentative agreement to participate in your proposal also lends a measure of credibility to your efforts. Advisors usually serve on a volunteer basis, but if it is allowed by your sponsor agency, you might consider writing in a budget line item to cover a small amount of basic communication or travel expenses. A cost-effective strategy is to plan your advisory board meeting(s) during a national conference where you and your advisors are already likely to be in attendance. If you are a student, form your board now, inviting scholars from inside and outside your home university, and establish a track record to support your career development goals.

Letters of support are also a good means of demonstrating the importance of your proposed research. Even if not required, a key endorsement from a leading scholar, a partnering agency, or other group can help your proposal stand out. To get these letters, be prepared to write each letter draft yourself and tailor every single one to each individual supporter. Do not use a form letter or you will receive five letters back with exactly the same wording. Do be sure that the letter refers to the exact, correct title of your project and is addressed to the right person or department at the funding organization or generally to the review committee. Alternatively, have the letters addressed to you.

A *research agenda* is a common way to demonstrate that you anticipate how this project contributes to a larger and broader knowledge-generating enterprise. By proposing to develop one during the course of your research project (and even better, in conjunction with your advisory board), you set yourself up for possible future funding opportunities. Later you can point to the research agenda as the source for new questions you wish to raise in subsequent proposals. (For an example of a research agenda publication, see AAG 2003.)

A *sustainability plan* is a tool to allow you to continue making progress beyond the grant period. Many granting agencies like to see that you are thinking about how to leverage their resources as a way of launching new, long-term efforts. Say a few words in your proposal about how you envision sustaining your work after the funds are gone (Carlson 2002, 47).

Matching funds or in-kind contributions are another way to demonstrate to funders that you will use their resources efficiently and stretch them to their highest impact. Even if not required, if acknowledgment of outside contributions is permitted, it is valuable to explicitly name what other resources you or your institution will bring to bear upon the effort, if nothing more than pointing to the use of facilities, extra faculty time spent working or writing on the project, and so forth.

References to your own publications in your proposal narrative and bibliography can help demonstrate your competence in the field and position yourself relative to other scholars. Having publications that relate to your proposed work certainly reinforces credentials in the eyes of proposal reviewers, but be sure that the references are relevant and appropriate to the project. Furthermore, even unfunded proposals themselves may later become a source of future articles, creating a synergy between your grant writing and scholarly publication writing activities.

Memorable project titles or acronyms can help reviewers recall your proposal over others, particularly in discussions among review committees. Proposal titles in fact make the first impression, so be careful to choose wisely, or the memory reviewers have may not be a positive one. For a thoughtful guide on prudent title selection, see Locke, Spirduso, and Silverman (2000, 126–29).

Dissemination activities include ways you will ensure that your research results are broadly known. Donors are often interested in investing in work that will be widely recognized and continue to have further impact on the scientific community or the public. It is easy to add a few strategic dissemination activities such as publishing your work in a journal, writing a newspaper op-ed about the results, holding a workshop or seminar or a panel at a conference to consider the results, creating a web site, or others (Box 11.11).

Evaluation processes are often required for major research projects, but even if yours is not mandatory, it is advisable to incorporate some kinds of evaluation procedures into your work. This ensures quality and effectiveness and allows you to map outcomes to original goals. It also provides excellent material for future proposals that build upon the same line of research. You may only need to include a few statements regarding how you plan to assess success. Or you may be required to contract with a professional evaluator that is external to the project. Formal, external evaluations

---

BOX 11.11

The following dissemination plan from Elwood (Appendix ii) demonstrates a commitment to both academic and public/community dissemination:

"I will disseminate results of the project in academic forums, the Humboldt Park community, and broader Chicago community development forums. Scholarly dissemination will take several forms and target several research areas in geography. I expect to produce multiple manuscripts for dissemination in urban geography, GIS, and geographic education journals. In the 2 years following completion of the project, I plan to produce a book focusing on the research results of the project, targeting the book toward an audience in urban geography, urban politics, community development. Throughout, I will make presentations at academic conferences focusing upon the project's research and educational initiatives. In terms of community dissemination, all data developed for the spatial analysis data library and materials produced in the spatial analysis projects will be available to the partner organizations, and will be disseminated in local community development and neighborhood organizing forums such as the Great Cities Conference – an annual gathering of scholars and practitioners to share results of research and action throughout Chicago. Data and results will also be shared with Humboldt Park's many other community organizations in presentations, and through the project website. This process will be facilitated by the ongoing partnership between DePaul's Egan Urban Center and the Humboldt Park community. I am particularly committed to this local dissemination of data and results, given the underdeveloped infrastructure of local support resources for community-based spatial analysis."

generally should be budgeted at about 10 percent of the overall budget (before overhead percentages are applied). Be sure that the evaluation utilizes a valid and reliable evaluation design, referring to methods that are appropriate to the kind of funding agency you are seeking support from. For example, foundations may prefer Logic Model Evaluations (Kellogg Foundation 2004); international organizations may require that you adhere to ISO standards (*e.g.,* ISO/IEC 19796-1:2005, the accepted reference criteria for evaluation of scenarios for Information and Communications Technology); or agencies such as the U.S. Department of Education may promote following any number of methodologies depending upon grade level, subject content, or other factors. The NSF makes available a number of guides that are helpful specifically for NSF-directed proposals (NSF 2007).

9. Achieve and Communicate Coherence

The whole is truly more than the sum of its parts. While writing each section of the narrative, completing the budget, putting together your biographical sketch, and compiling all of the other parts of the proposal, you should strive for coherence among all of the elements. While you will not be able to eliminate all possible contradictions from any text, the idea is to create a complete package that minimizes a sense of the disjoined. Don't leave this to chance: systematically structure your proposal to generate this coherence by using a matrix or other device that helps you to question the relationships among elements.

For example, following line A as marked in Figure 11.1: Do your research objectives actually address the problem or need you identify? Line

FIGURE 11.1 A coherence matrix for research proposals.

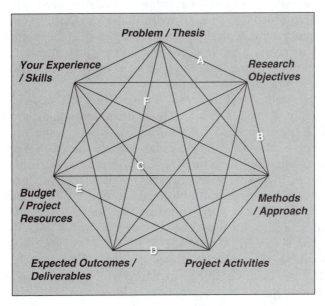

B: Will your methodology really enable you to answer the question(s) you pose? Line C: Do your skills match up with the demands of the research activities that you propose? Line D: Will the project activities suffice to produce the deliverables that you promise? Line E: Do budget items line up with the activities? Line F: Will your expected results actually inform the research context? Do all of these elements overlap with the goals and perspective of the sponsor organization? And so forth (see Box 11.12 and also the example in Appendix vii).

You can create your own matrix using required elements of the specific call for proposals you are responding to. Or, you can create a general matrix tailored to your research by incorporating your own research questions. Activity 11.2 outlines a procedure for how you might create and use this kind of tool not only to design your project, and in turn structure your proposal, but also to review written drafts for completeness and coherence.

10. Live It

Preparing competitive research grant proposals is just as much about preparing yourself for a successful career as it is about getting resources for one particular project. Writing grants successfully is a continual process that is best practiced as integral to your professional life—not just parallel to it. Here I highlight a few simple organizational tools you can employ to create a constant state of readiness that will allow you to respond to funding opportunities as a matter of professional practice, rather than approaching it as a project (or chore!) in itself.

- Keep files/folders (either digital or hard copy) on possible research questions or problems that interest you. A research idea diary can help you focus on lines of work that you would like to elaborate, without having to flesh them out completely in the moment. You will be surprised in reviewing past entries just how your thoughts develop over time. This can then be a source for responding quickly to new funding opportunities.
- Keep similar but separate files/folders on possible sponsors amenable to your work. Organize this information, as you gather it into a spreadsheet so that you can find what you are looking for at a glance and you can see relationships between opportunities. Information on these funding agencies can come from a variety of sources, including your university's SRO. You can also simply use any search engine on the Internet combining key words for your research and the word "funding" or "grant" to dig up potential opportunities. Or mine the CVs of scholars in your field, as suggested above in "Create the Right Fit." Be sure the spreadsheet includes a column for deadlines; plan ahead, and prioritize.
- When you have identified a match between a sponsor and one of your research questions or when you have a sufficiently developed

BOX 11.12

With limited space in the proposal narrative, you can create a sense of coherence by lining up goals/purpose, activities, the timeline, and deliverables as shown in this example table (Solís, Appendix vii):

Category	ACTIVITY	PURPOSE	M	A	M	J	J	A	S	O	N	D	J	F	Deliverable
Regional Status	Initial Advisory Board & Planning Session at AAG Annual Meeting in Chicago	To review research plan, identify additional collaborations	X												
	Gather departmental data	To inventory geography in the Americas	X	X	X	X	X	X							
	Develop survey respondent listings	To identify IRC activities and participants			X										
	Draft and pilot survey instruments; translate				X										
	Implement surveys							X							
	Mail reminders to non-respondents	To increase response rates									X				
	Conduct follow-up telephone interviews to non-respondents										X				
	Create and input data into GIS	To visualize and understand spatial patterns						X	X	X	X	X	X	X	
	Analyze data (spatial and statistical)									X	X	X			
	Publish data in Directory and online	To share information and facilitate linkages										X	X	X	Directory of Geography in the Americas
Local Contextual Dynamics	Research local case study information sources	To discover and understand qualitative context of IRC	X	X	X										
	Participant observation at CoK		X	X	X	X	X	X	X	X	X	X	X		
	Investigate disciplinary history				X	X									
	Devise focus group methodology participants						X	X							
	Conduct focus groups							X	X						
	Data analysis of focus groups								X						
	Devise interview methodology respondents	To test assumptions, determine extend or quantify findings								X					
	Conduct interviews										X				
	Data analysis of interviews										X				Case Study
Broad Facilitation Assessment	Research presented at US geography conference	To share results and gather geography community feedback, seek input for recommendations								X					
	Plan Geography Summit									X	X				
	Send invitations for Geography Summit and Advisory Board meeting										X	X			
	Conduct effectiveness model evaluation analysis	To develop recommendations										X	X		
	Write project reports / NSF report	To determinate results and catalyze future IRC										X	X		Set of Best practices; Sustainability Plan
	Hold Geography Summit												X		Geography Summit
	Dissemination to geography community and other disciplinary societies													X	International Model

research question arising in response to a particular call for proposals, you should launch a special writing operation. Start with a proposal writing timeline for each proposal working backward from the deadline, and identify key dates for finishing certain benchmarks.

• Write for fifteen minutes a day. Whether or not you have a particular funding opportunity, call for proposals, or research sponsor in mind, your writing skills to address this special audience will develop by frequent practice. Successful grant seekers almost always have one or more proposals in development, with draft documents specifically designed to capture ideas, arguments, and thoughts as they occur saved on their hard drives. This reduces the need to create the research proposal completely from scratch once you find a suitable sponsor organization. This reinforces the advice from Chapter 1 by Ken Foote on time management, that points out how people working in shorter, more frequent periods are better able to sustain momentum and continue progress as opposed to "bingeing" around deadlines. This approach is not only more productive but also less stressful.

Finally, as Iain Hay and Mark Israel discuss in the next chapter, there are clear ethical considerations that must be taken into account when writing grants. These are familiar refrains against plagiarism, integrity, honesty, informed consent, reciprocity, conflicts of interest, and other stances appropriate for aspiring scholars to live up to. (See Locke, Spirduso, and Silverman 2000; Chapin 2004.)

NOW WHAT?

It can be an exhilarating moment to finally submit a grant proposal for consideration after all of the hard work involved in preparing it. However, soon the realization hits that you will have to wait sometimes as long as six months for a decision. At these moments, turn to the next proposal and do not continually ask for updates from the program officers. Reject the temptation to make grammatical corrections or substantive updates and resend. Meanwhile, realistically prepare yourself for what to expect next: actual funding rates are indeed quite low, and even experienced grant writers can expect success rates on the order of 30 percent or less. Instead, focus on the other benefits that the writing process brought you. Reflect on what you might do better next time, regardless of whether you receive a positive or negative response for funding.

After proposals are submitted, they enter some kind of formal review process. Often this includes peer review, but for some foundations, the staff or a board of directors considers the applications. In any case, once you receive notification of a decision, you will usually also receive copies of the feedback from reviewers. If you do not, you should certainly request them. In the event of a declined proposal (a likelihood given the competitiveness of many programs), looking carefully through the feedback will help you understand that the process itself is instructive. A rejection of a proposal does not mean your ideas are not worthy of future consideration, it may simply indicate temporarily limited funding or unusually heavy competition. Read all reviews and use them to revise and resubmit rejected proposals or to reformulate the next one (Ogden and Goldberg 2002, 181) (Box 11.13). Even the reviews of accepted proposals should be carefully considered, because they often include very helpful comments that will improve your funded project.

BOX 11.13

Kenney and Patton (Appendix iii) revised and resubmitted their proposal three times to NSF before receiving the award. Note that the overall rating for funding priority in the unfunded proposal sample is the same as the proposal that was ultimately awarded (medium priority), although considered by two different panels. There are discrepancies or disagreements among individual reviewers that are important to consider together. It is worth comparing the proposals, especially the summaries, to see how the authors utilized feedback constructively to craft a winning proposal. In particular, look at the difference in the opening paragraphs in response to reviewers' comments, reflecting a transformation from a project focused primarily on database construction to one that has the analysis front and center:

(unfunded proposal text):

> "This study examines the spatial relationships between successful startup firms and various constituents of a startup's institutional support network that contribute to its birth and growth. In particular, this project requests funding to extend an existing database to include data on all domestic firms that went public between June 1996 through 2000. The existing database includes geographic information on the firms themselves, and five members of their support network: lawyers, venture capitalists, investment bankers, advisors in terms of members of the board of directors, and the universities that trained the management team of these firms."

(review excerpts):

> "It is unclear if the study will result in theoretical verification since few details on how the hypotheses will be tested are provided . . . It is primarily focused on the collection of data (although a number of hypotheses are laid out it is not clear how the researcher will proceed in answering them) . . . it is primarily a database building proposal which the author plans to share with the research community . . . the major weakness of this proposal is the lack of analysis. Though hypotheses were described, no information was provided to explain how the data will be analyzed and how the hypotheses will be tested. The methodology section is just the method to complete the database."

(revised, funded proposal text):

> "The role of social actors supporting entrepreneurship has recently been recognized as critical, yet their geography is not well understood. This is true despite the fact that the literature has recognized the importance of entrepreneurship for the birth and growth of industrial clusters. Drawing upon the results of research in high-technology clusters, there has been a general assumption that supporting actors, such as lawyers, venture capitalists, and others, should be located in close proximity to the entrepreneurs because of the necessity for face-to-face interaction to transmit tacit knowledge. More recently, some spatial scientists have remarked that even in highly clustered industries characterized internally by "local buzz," "global pipelines" to actors outside the cluster exist and are important. The proximity and role of these support network actors across industries and regions will be explored through testing a variety of propositions derived from the literature on clusters and entrepreneurship."

In conclusion, the entire process of writing research grant proposals—from developing ideas to writing convincing text to considering reviewers' feedback—can be conducted in a way that solidifies your research ideas and shapes your academic career in positive directions. Integrating these practical recommendations and habits of mind and action into your professional life can help you prepare yourself as a successful researcher *and* grant winner.

Additional Resources

Abler, R. 1989. How to win extramural research awards. In *On becoming a professional geographer*, ed. M. Kenzer, 170–82. Caldwell, NJ: Blackburn Press.

> Although written some time ago, Dr. Abler's advice is salient and timeless. Coming from this renowned geographer, in an elegant and witty style, the text itself is a good example of clear and persuasive writing. A short and easy piece to cover a good deal of conceptual ground.

Chapin, P. G. 2004. *Research projects and research proposals: A guide for scientists seeking funding*. Cambridge, MA: Cambridge University Press.

> A more recent and comprehensive treatment written by a former NSF program officer is right on target for an aspiring academic audience, which is likely to try a submission to this funding agency. Chapin's approach is to contextualize the process of proposal writing within the broader practice of planning and implementing a research project. Beyond its clear guidance on designing and writing the proposal, it includes discussion of managing awarded projects, ethical responsibility, and dealing with proposal rejection.

Geever, J. C. 2001. *The Foundation Center's guide to proposal writing*, 3rd ed. New York: The Foundation Center.
> A widely read classic and indispensable general resource for proposal writing, especially for applied projects, but also appropriate for studies. The perspective is particularly valuable for researchers seeking to submit proposals to private foundations or donors that are nontraditional in terms of academic research proposals.

Locke, L. F., W. W. Spirduso, and S. J. Silverman. 1999. *Proposals that work: A guide for planning dissertations and grant proposals*. Thousand Oaks, CA: Sage Publications.
> For graduate students, this book not only reliably advises on key aspects of developing a research thesis but also navigates through the dissertation proposal process itself, including how to form a dissertation committee and make the oral presentation. Its organization is particularly creative around nine functions that proposals must perform. The sections on writing a literature review, designing budgets, and discussing qualitative research will make this a valuable reference work in any academician's library.

Hall, M., and S. Howlett. 2003. *Getting funded: The complete guide to writing grant proposals*, 4th ed. Portland, OR: Portland State University Press.
> Although written for a much broader audience than just academic researchers, readers will find plenty of practical value in this volume. This edition includes a section for instructors teaching proposal writing, including a sample syllabus for eleven-week and three-week courses, and suggested assignments related to each chapter.

References

Abler, R. 1989. How to win extramural research awards. In *On becoming a professional geographer*, ed. M. Kenzer, 170–82. Caldwell, NJ: Blackburn Press.
Association of American Geographers (AAG). 2003. *Geographic dimensions of terrorism: A research agenda*. Washington, DC: AAG Booklet Publication.
———. Various years. *Guide to geography programs in the Americas*. Washington, DC: AAG Booklet Publication.
Carlson, M. 2002. *Winning grants: Step by step*, 2nd ed. San Francisco: Wiley.
Catalog of Federal Domestic Assistance (CFDA). n.d. Developing and writing grant proposals http://12.46.245.173/pls/portal30/CATALOG.GRANT_PROPOSAL_DYN.show (last accessed 4 February 2008).
Chapin, P. 2004. *Research projects and research proposals: A guide for scientists seeking funding*. Cambridge, MA: Cambridge University Press.
The Foundation Center. 2006. Proposal budgeting basics. http://foundationcenter.org/getstarted/tutorials/prop_budgt/index.html (last accessed 4 February 2008).
Geever, J. 2001. *The Foundation Center's guide to proposal writing*, 3rd ed. New York: The Foundation Center.
Gerin, W. 2006. *Writing the NIH grant proposal: A step-by-step guide*. Thousand Oaks, CA: Sage Publications.
Gewin, V. 2004. Mapping opportunities. *Nature* 427 (22): 376–77.
Henson, K. 2003. *Grant writing in higher education: A step-by-step guide*. Boston, MA: Allyn & Bacon.

Jackson, D. 2005. The insular American. *The Boston Globe*, 29 January.

Kellogg Foundation. 2004. *Logic model development guide: Using logic models to bring together planning, evaluation, and action.* Publication #1209. Battle Creek, MI: Kellogg Foundation.

Krathwohl, D. 1988. *How to prepare a research proposal: Guidelines for funding and dissertations in the social and behavioral sciences*, 3rd ed. Syracuse, NY: Syracuse University Press.

Lee, J., and S. Clery. 2004. Key trends in higher education. *American Academic*1 (1): 21–36.

Locke, L., W. Spirduso, and S. Silverman. 2000. *Proposals that work: A guide for planning dissertations and grant proposals.* Thousand Oaks, CA: Sage Publications.

National Center for Educational Statistics (NCES). Various Years. *Digest of education statistics.* Washington, DC: NCES.

National Research Council (NRC). 1997. *Rediscovering geography: New relevance for science and society.* Washington, DC: National Academies Press.

———. 2006. *Beyond mapping: Meeting national needs through enhanced geographic information science.* Washington, DC: National Academies Press.

National Science Foundation (NSF). 2005. A guide for proposal writing. http://www.nsf.gov/pubs/2004/nsf04016/start.htm (last accessed 4 February 2008).

———. 2007. Resources for NSF project evaluation. http://www.ehr.nsf.gov/rec/programs/evaluation/nsfresources.asp (last accessed 4 February 2008).

Ogden, T., and I. Goldberg, eds. 2002. *Research proposals: A guide to success*, 3rd ed. San Diego, CA: Elsevier Science.

Pandit, K. 2004. Geography's human resources over the past half-century. *The Professional Geographer*56 (1): 12–21.

Pfirman, S., and the AC-ERE. 2003. *Complex environmental systems: Synthesis for earth, life and society in the 21st century: A 10-year outlook in environmental research and education for the National Science Foundation.* Washington, DC: National Science Foundation.

Przeworski, A., and F. Salomon. 1995. The art of writing proposals: Some candid suggestions for applicants to Social Science Research Council competitions http://www.ssrc.org/fellowships/art_of_writing_proposals.page (last accessed 4 February 2008).

Richardson, D., and P. Solís. 2004. Confronted by insurmountable opportunities: Geography in society at the AAG's centennial. *The Professional Geographer* 56 (1): 4–11.

Schumacher, Dorin. 1992. *Get Funded! A Practical Guide for Scholars Seeking Research Support from Business.* Newbury Park: Sage Publications.

Solem, M. 2004. Center for the Advancement of Geography Education. Unpublished proposal manuscript to the National Science Foundation.

Watts, M. 2001. Conceptualizing, writing, and revising a social science research proposal. http://globetrotter.berkeley.edu/DissPropWorkshop/sitedescription.html (last accessed 4 February 2008).

CHAPTER 12

Private People, Secret Places: Ethical Research in Practice

Iain Hay and Mark Israel

There are at least five reasons why college and university faculty do—and should—take research ethics seriously. These include

- to protect others and minimize harm to people and places;
- to maintain and assure the public trust that allows continued useful research;
- to ensure research integrity (thereby maintaining faith in one another's research outcomes);
- to satisfy organizational and professional requirements and expectations (*e.g.*, universities, professional bodies); and
- to allow us to cope with new and more challenging problems.

And, as Israel and Hay (2006) have argued, there is value in developing ethical prowess to help ensure your work is not disabled by biomedically led regulation of ethical practice uninformed by geographical research practices. Sadly, however, it can be more difficult to behave ethically than you might like because many of us do not (1) have the philosophical training to negotiate sometimes difficult ethical terrain; (2) recognize ethical challenges when they appear; (3) have the time to make the best decisions; or (4) because we have not anticipated problems that may arise in our work. In this short chapter (and two associated activities—Activity 12.1: Research Ethics: The Regulatory Context and Activity 12.2: A Case for Ethics), we take some small steps to help you address these ethical issues.

FROM REGULATORY COMPLIANCE TO ETHICAL CONDUCT

Researchers are expected to comply with various professional ethical codes and government regulations while undertaking research. Among the most significant of these in the U.S. are those associated with the "Common Rule." Since 1991 a common federal policy (Title 45 of the Code of Federal Regulations Part 46, Subpart A) has set out regulatory procedures for all research involving human subjects. The Common Rule is now followed by at least seventeen federal agencies including the National Science Foundation (NSF), the Environmental Protection Agency, and the Agency for International Development (see Department of Health and Human Services [DHHS] 2007). American research institutions such as universities must comply with the Common Rule for *all* research if they are to remain eligible for funding provided by government agencies subscribing to the Rule. Generally, this is achieved by Institutional Review Board (IRB) scrutiny of research proposals and activity. So, for example, U.S. federal agencies such as the NSF may require that a research proposal provides evidence of examination by an IRB that is registered with DHHS' Office for Human Research Protection (OHRP) and has oversight of the researcher's activities. (For additional information about the role of IRBs in proposal development, please see Chapter 11, "Preparing Competitive Research Grant Proposals.")

In Canada, geographers need to consult the Tri-Council Policy Statement (2003), which sets out national norms that are applied at a local level through a review process conducted by multidisciplinary, independent Research Ethics Boards (REBs) (Tri-Council 1998, Article 1.1). With few exceptions, all proposed research involving living human subjects must be scrutinized prospectively by a REB.

In addition to national regulatory requirements, geographers should consider guidelines set out by relevant professional bodies such as the Association of American Geographers (1998) and the American Association of University Professors (1987). There are other, more specialist bodies in the U.S. that also provide guidance to their members—these include the Academy of Board Certified Environmental Professionals (2006), American Society for Photogrammetry and Remote Sensing (2006), GIS Certification Institute (c.2003), and the Urban and Regional Information Systems Association (2003).The Illinois Institute of Technology's Center for the Study of Ethics in the Professions (2007) sets out a collection of more than 850 professional and institutional codes that provides a helpful starting point for finding specialist guidance. In short, before you commence any research work, it is important to be aware of and respond to relevant institutional and professional regulations. Activity 12.1 encourages you to explore those regulations.

Research ethics guidelines typically respond to the concerns and ambitions outlined in the bullet points at the start of this chapter. Despite their good intentions, it is worth noting however that these processes are some-times argued to do more harm than good. For example, they are said to dis-courage thoughtful ethical reflection, promote excessive interference in research designs, cause unnecessary research delays, and overemphasize formal consent procedures (Citro, Ilgen, and Marrett 2003; Bosk and De Vries 2004). However, the consequences of failing to work within existing regula-tory frameworks can be severe, for individuals and their institutions. For instance, the OHRP has suspended (or threatened to suspend) the federally funded research of entire institutions where there has been failure to comply with federal research ethics regulations (Oakes 2002; Sieber, Plattner, and Rubin 2002). And academics as senior as university presidents have lost their jobs for ethical infringements (see Israel and Hay 2006, 116–17). Not surpris-ingly, then, it follows that at the institutional level most attention is given to how researchers can best respond to the needs of regulators (Oakes 2002; Bach 2005) or to legal issues relating to research (Boruch and Cecil 1983; Chalmers and Israel 2005).

Having acknowledged the regulatory context of research, we wish to focus the remainder of this chapter on the ethical aspects of day-to-day conduct of research, where the ethical challenges you face may not have been anticipated, or do not offer the luxury of time for consultation of rules and guidelines, or do not allow you to confer with colleagues (see Box 12.1). You need to be prepared for these occasions and able to call upon the knowledge and experience necessary to avoid ill-judged behavior or disastrous consequences. In what remains of this chapter and in Activity 12.2, we draw from our recent book (Israel and Hay 2006) to focus on how you might engage your "moral imagination" (Hay 1998), honing the skills and perspectives needed to deal with unanticipated or difficult ethical challenges.

BOX 12.1 *An Ethically Important Moment*

Picture this scene. You are a researcher working on a study examining women's experiences of heart disease. You are interviewing Sonia, a woman in her late 40s with diagnosed heart disease. . . . The interview is progressing well. Over a cup of tea in Sonia's kitchen, you inquire about the impact of heart disease on her life. Sonia stops and closes her eyes. After a few moments' silence, you notice her eyes welling up with tears. Sonia tells you she is not coping—not because of her heart disease, but because she has just found out that her husband has been sexually abusing her daughter . . . (Guillemin and Gillam 2004, 261)

How would you respond? Would you continue with the interview? Would you offer Sonia any support? Do you have any legal or moral responsibility to report this infor-mation to relevant authorities?

HOW CAN RESEARCHERS DECIDE WHAT TO DO WHEN PRESENTED WITH AN ETHICAL DILEMMA?

Researchers often base ethical practice on one of the three principles set out in the landmark Belmont Report (National Commission for the Protection of Human Subjects of Biomedical and Behavioral Research 1978)—justice, beneficence, and respect for persons. However, sometimes these three principles come into conflict, and it is necessary to return to fundamental approaches to resolve an ethical dilemma.

There are two key normative approaches to ethical practice (see Israel and Hay 2006). The *consequentalist* approach focuses on the practical consequences of actions—whether or not an action is ethical depends on its associated costs and rewards. By contrast, *deontological* approaches reject the emphasis on consequences and suggest instead that certain acts are seen as good in themselves and must be viewed as morally correct because, for example, they uphold promises, demonstrate gratitude, or show loyalty. It is possible, therefore, for something to be ethically correct even if it does not promote the greatest balance of good over evil. These two positions underpin the strategy for coping with ethical dilemmas developed in Hay (2003) and Israel and Hay (2006) and are summarized in Table 12.1.

Using such a strategy to work thoughtfully through the problem, you may reach a considered and defensible position. If your research situation provides the time and opportunity, it can also be useful to discuss ethical problems (actual or prospective) with colleagues—where and when appropriate—and, if possible, more extensively with those people involved in the research (Fetterman 1983).

Below, we analyze one dilemma "Private People" (Box 12.2) confronted by a geography research student (based on a true case; individual and organizational names have been altered) and, to encourage you to think through parallel issues in another quite different case, we present another scenario, "Secret Places" (Box 12.3). As you continue reading, apply the steps to resolving an ethical dilemma from Table 12.1 to "Secret Places," drawing from our discussion of "Private People" for insights. This process mirrors the procedures set out in Activity 12.2.

TABLE 12.1   Steps to resolving an ethical dilemma.

- Identify the issues, identify the parties
- Identify options
- Consider consequences
  - *Who or what will be helped?*
  - *Who or what will be hurt?*
  - *What kinds of benefits or harms are involved, and what are their relative values?*
  - *What are the short- and long-term implications?*
- Analyze options in terms of moral principles
- Make your own decision and act with commitment
- Evaluate the system
- Evaluate yourself

Adapted from Israel and Hay (2006, 132).

Identify the Issues, Identify the Parties

Not surprisingly, the first steps in ethical decision-making are to identify the existence and nature of the ethical problem and identify the range of different stakeholders involved: who will be affected, what is at stake for them? It is helpful to think of stakeholders as concentric and increasingly larger groupings starting with those immediately affected by a situation or decision; then moving out to relevant institutions (for example, university, employer); to communities of social science researchers; and finally to society as a whole (Bebeau et al. 1995).

BOX 12.2 *Private People*

Margaret recently completed research that involved distributing a confidential questionnaire to members of an environmental nongovernment organization, "The Gray Greenies." Members of this organization are mainly elderly environmental activists. Margaret works part-time for the organization and feels that her employment tenure is somewhat precarious. The research project was assessed formally by her academic supervisors and Margaret plans to present a modified report to the NGO. While leaving work one day, Margaret meets one of the organization's members, Montgomery, who asks what the study results are. Margaret outlines some of the findings, such as the percentage of the membership who are having trouble, due to old age and declining health, performing their voluntary duties as vehicle drivers for the organization. Montgomery asks Margaret to identify those members who are having difficulties. It is possible that he could use the information to develop strategies to help those members. He might also campaign to have the same people redirected to less demanding but also less fulfilling volunteer roles.

(Adapted from Hay 2003; Israel and Hay 2006).

BOX 12.3 *Secret Places*

Dr Lachlan Driver is a British geographer who is using geographical information systems and multi-criteria modeling techniques to map the wilderness continuum in the U.K. His intent is to identify the wildest and most remote areas in the U.K., many of which are in Scotland. At the time of Driver's research, there are plans afoot to extend the National Park system of England and Wales into Scotland. If Dr Driver publicly identifies the many wilderness areas (particularly in Scotland) that are presently unprotected they may be formally protected under National Parks legislation. But, in doing this, he will simultaneously be drawing the areas to the attention of the recreational lobby looking for wilderness experiences. If the work is not made public, there is a chance that the areas may go unprotected when park boundaries are formalized.

(Adapted from Hay and Foley 1998).

In "Private People," stakeholders include the research student, the members of the NGO, and the people they assist. They also include other researchers and those members of the general public who might be put at physical risk by elderly volunteers. The relationships are complex. For example, Margaret has a confidential research relationship with the volunteers but also depends on the organization for her job. Her tenure in that role may be influenced by Montgomery, who may indeed be looking out for the best interests of the volunteers and those whose needs they serve.

Identify Options

There may be several plausible responses to an ethical problem, and researchers should not discard possibilities prematurely. In some instances, options may be shaped by organizational and institutional codes of ethics, or by relevant laws. For example, as we saw in Box 12.1, what might the researcher do after being told of child sexual abuse? This is what Palys and Lowman (2001) termed "heinous discovery"—when researchers discover disturbing material they could not have anticipated. It is improbable that a researcher's procedural ethics will have anticipated these possibilities, and legislation surrounding mandatory reporting of child sex abuse or intention to commit a crime may require official reporting of each interviewee's revelation. Arguably, the researcher is faced with two options: honor the promise of confidentiality to the interviewee or abide by laws and reveal the incident to relevant agencies.

Cases of heinous discovery are uncommon and the challenges you will face in your everyday research may be less confronting and less obvious. For instance, in "Private People" (Box 12.2), Margaret has at least the following set of alternatives open to her. She can (1) spill the beans by revealing to Montgomery the names of those members of The Gray Greenies who are experiencing difficulties with their duties; (2) keep quiet and maintain confidentiality; (3) bite the bullet and confront Montgomery about the difficult position in which he has placed her; or (4) pass the buck and advise the NGO's director she has been approached with a request for information given in confidence. She will also have to decide in this option whether or not to divulge Montgomery's name as the person making the demand. However, passing the buck to the director does not actually resolve the dilemma: it simply offloads some responsibility to another person.

Consider Consequences

With a clear appreciation of issues, parties, and options, you can pursue a consequentalist approach to the dilemma, considering the range of positive and negative outcomes associated with each option. This can be difficult when it is unclear what the consequences of any actions might be (Sieber 1982) and it may be tempting to procrastinate, jump straight to an ill-considered decision, or move immediately to decision making based on consideration of moral principles such as honesty and equality. We think it more sensible to gather information on the harm and benefit that may be caused by each option, focusing on those with the highest probability of occurring. The issues of benefit maximization and harm minimization can be organized around four questions: who or what will be helped?; who or what will be hurt?; what kinds of benefits and harms are involved, and what are their relative values?; and what are the short- and long-term implications? We can deal with these in turn.

Who or what will be helped? Option 1 for Margaret—revealing the names of helpers experiencing difficulties—might see volunteers forced into less fulfilling roles or alternatively mechanisms established to assist members of the organization facing difficulties. In either case, and if they become aware of her revelations, people who provided Margaret with information in confidence could take actions to have her dismissed or call her research into question. It is possible that Option 2—maintaining confidentiality in the face of a request from a member—or Option 3—confronting a member—could result in work problems for Margaret and perhaps her eventual dismissal. It might also mean that some of the more infirm volunteers could continue to endanger themselves and others in the conduct of their duties. If Margaret selects Option 4 and advises the director of the request for information, he may ask for Montgomery's name. If Margaret provides it, she will probably have made an enemy in Montgomery and that might jeopardize the reception members give her research or her continuing employment prospects with the organization. If she refuses to divulge Montgomery's name, the director may disregard her concerns, feeling that he cannot act. Margaret may then have to deal with Montgomery on her own. It is also possible that the director will see the justification in Montgomery's request for volunteers' names and make the same difficult request of Margaret.

Who or what will be hurt? Option 1 could see Margaret hurt through the loss of her job, and volunteers hurt both through reassignment to work they enjoy less and through Margaret's breach of confidence. Options 2 and 3 could also result in Margaret losing her job. It might also end with volunteers and members of the public injured or killed. Option 4 could result in an angry Montgomery being reprimanded by the NGO director. It could also eventually yield the same harms set out in Options 1 and 2 if the director also asks Margaret for the names of those volunteers having problems in their current roles.

What kinds of benefits and harms are involved, and what are their relative values? You may view some things such as healthy bodies and clean beaches as more valuable than others, such as new cars. Some harms such as violation of trust may be more significant than others like lying in a public meeting to protect a seal colony or gecko preserve. It is important then to give "weights" to the various costs and benefits of options presented. In The Gray Greenies case, you might regard the prospects of people being killed in accidents involving volunteers as of greater importance than the prospects of Margaret losing her job or Montgomery feeling upset.

What are the short- and long-term implications?   Would The Gray Greenies survive the bad press that might follow from any of its volunteers being involved in a major motor vehicle accident? What might be the effects of such a tragic incident for Margaret herself? How much does Margaret's current job mean to her? Of what significance is a breach of confidence by Margaret for the overall credibility and reputations of other social science researchers? As you adopt long-term time-scales in your consideration of implications, you must deal with an increasingly uncertain range of futures. Moreover, it may be difficult to assess the relative importance of immediate possibilities against more distant ones (MacIntyre 1985).

These four questions structure a consideration of options, providing a useful viewpoint to *one* side of an ethical problem. We now move on to consider matters from a deontological position.

Analyze Options in Terms of Moral Principles

The researcher faced with an ethical dilemma needs to consider how options measure up against moral principles such as honesty, trust, individual autonomy, fairness, equality, and recognition of social and environmental vulnerability. Are any of these being violated? How and for whom? In some instances, some principles may be regarded as more important than others. For instance, in "Private People," the promise of confidentiality Margaret made to volunteers might override other matters. Where appropriate and if time is available, researchers may get valuable insights by consulting with colleagues, other researchers, counselors, or people outside the research context with specialist ethical/moral training. These people may be able to help clarify principles and ideas. Ultimately, however, the decision is the responsibility of the researcher.

Make Your Own Decision and Act with Commitment

It is now possible for the researcher to bring together analyses of consequences and principles and make an informed and thoroughly considered decision. This is not easy. No single decision is likely to allow any of us to sidestep adverse consequences or violated principles. So, you may find yourself forced to choose the lesser of several evils. Such disagreeable decision-making can be made a little easier. First, identify exactly where the problem lies—for example, the point at which some value such as fairness has been violated. Second, clarify the nature of the value conflict through analogies. How have other similar cases been handled and what were the outcomes? Is the issue you are investigating similar to or different from those analogies

(MacDonald 2002)? What would happen if certain elements or individuals in the scenario were changed or if the stakes were higher or lower? For example, how would you react if Margaret's information had been about respondents' illicit drug-taking activities? Third, following the advice presented earlier, outline and substantiate the desirable or undesirable consequences of each possible position. Finally, set priorities. Given the consequences, which value or values should take priority?

Having made these deliberations, but before committing to a course of action, you may find it useful to reflect on the following questions (adapted from MacDonald 2002): Will I feel comfortable telling a close family member such as my mother or father what I have done?; How would I feel if my colleagues knew what I was doing?; Would I be embarrassed if my actions were to attract media attention?; How would I feel if my children followed my example?; Is this the behavior of a wise and virtuous person?

You can now act, having reached a well-considered decision, and on the understanding that the decision is your responsibility and no one else's. As Guillemin and Gillam (2004, 269) remind us, "Ultimately, responsibility falls back to the researchers themselves—they are the ones on whom the conduct of ethical research depends."

Evaluate the System, Evaluate Yourself

Finally, it is important to reflect on the circumstances that led to the dilemma or problem (Hertz 1996). In some cases it may be possible to resolve systemic difficulties by devising new working practices—institutional or personal. Thus, your "ethics in practice" may become part of more formal procedural ethics. From Margaret's experience, you might conclude, for example, that unless she was prepared to deal with the recriminations of volunteers removed from duties, she should not have asked them any questions about their physical abilities to complete their work. In other words, she—and indeed, her Institutional Review Board (or Research Ethics Board in Canada)—may not have thought carefully enough about the potential implications of her results when she was designing her study. You might also conclude that given the nature of Margaret's employment the problem of conflict of interest needed to be tackled before the research started. It was partly Margaret's precarious job tenure that made her vulnerable to pressure from Montgomery.

As an individual researcher, consider how you might respond to similar problems in the future. For example, after being advised by an interviewee of a case of incest (Box 12.1), Sonia might consider how she (and others) could deal with unanticipated personal disclosures in future research. What skills does she need to develop? What resources might be useful?

Reflection and revision are vital to the development of good, ethical research practice. They may not prevent you from repeating the mistakes of the past, but without procedural and personal change based on experience and thoughtful contemplation we are all condemned to repeating our mistakes.

CONCLUSION

Ethical practice involves more than responding to the questions and demands of professional bodies, IRBs, and REBs. It requires a commitment to ethical conduct and the ability to cope with dilemmas as they arise in everyday practice—during an interview, in a parking lot, or at a research meeting. Unquestionably, some of these challenges will catch you flat-footed. Others may allow you the luxury of time for thoughtful, structured reflection of the type set out in this chapter. No matter how they present themselves—be they private people or secret places—each dilemma offers you an opportunity to reflect on and improve your practices, your skills, and the contexts within which we all operate. To that end, we suggest that all researchers need to extend their ethical skills and knowledge, not only to help resolve difficult, day-to-day dilemmas in research, but also to enable our constructive engagement with local and national regulatory bodies.

ACKNOWLEDGMENTS

We would like to thank Eric Compas, Richard Reddy, Michael Solem, and several anonymous referees for their helpful comments on an earlier draft.

References

Academy of Board Certified Environmental Professionals (ABCEP). 2006. Code of ethics and standards of practice for environmental professionals. http://www.abcep.org/ (last accessed 29 August 2007).

American Association of University Professors (AAUP). 1987. Statement of professional ethics. http://www.aaup.org/aaup (last accessed 29 August 2007).

American Society for Photogrammetry and Remote Sensing (ASPRS). 2006. Code of ethics. http://www.asprs.org/membership/certification/appendix_a.html (last accessed 29 August 2007).

Association of American Geographers. 1998. *Statement on Professional Ethics*. http://www.aag.org/Publications/EthicsStatement.html (last accessed 29 August 2007).

Bach, B. W. 2005. The organizational tension of othering. *Journal of Applied Communication Research* 33 (3): 258–68.

Bebeau, M. J. 1995. Developing a well-reasoned response to a moral problem in scientific research. In *Moral reasoning in scientific research. Cases for teaching and assessment*, eds. M. J. Bebeau, K. D. Pimple, K. M. T. Muskavitch, S. L. Borden, and D. H. Smith. Proceedings of a workshop at Indiana University, Poynter Center for the Study of Ethics and American Institutions, Indiana University, Bloomington.

Bebeau, M. J., K. D. Pimple, K. M. T. Muskavitch, S. L. Borden, and D. H. Smith, eds. 1995. *Moral reasoning in scientific research. Cases for teaching and assessment.*

Proceedings of a workshop at Indiana University, Poynter Center for the Study of Ethics and American Institutions, Indiana University, Bloomington.

Boruch, R. F., and J. S. Cecil, eds. 1983. *Solutions to ethical and legal problems in social research.* New York: Academic Press.

Bosk, C. L., and R. G. De Vries. 2004. Bureaucracies and mass deception: Institutional Review Boards and the ethics of ethnographic research. *Annals of the American Academy of Political and Social Science* 595 (September): 249–63.

Chalmers, R., and M. Israel. 2005. *Caring for data: Law, professional codes and the negotiation of confidentiality in Australian criminological research.* Canberra, Australia: Criminology Research Council. http://www.aic.gov.au/crc/reports/200304-09.html (last accessed 29 August 2007).

Citro, C. F., D. R. Ilgen, and C. B. Marrett. 2003. *Protecting participants and facilitating social and behavioral sciences research.* Washington, DC: The National Academies Press.

Department of Health and Human Services. 2007. Office for Human Research Protections (OHRP) IRB guidebook. http://www.hhs.gov/ohrp/irb/irb_guidebook.htm (last accessed 18 July 2007).

Fetterman, D. M. 1983. Guilty knowledge, dirty hands, and other ethical dilemmas: The hazards of contract research. *Human Organization* 42 (3): 214–24.

GIS Certification Institute c.2003. Code of ethics. http://www.gisci.org/code_of_ethics.aspx#_edn1 (last accessed 29 August 2007).

Guillemin, M., and L. Gillam. 2004. Ethics, reflexivity and "ethically important moments" in research. *Qualitative Inquiry* 10 (2): 261–80.

Hay, I. 1998. Making moral imaginations. Research ethics, pedagogy, and professional human geography. *Ethics, Place and Environment* 1 (1): 55–75.

———. 2003. Ethical practice in geographical research. In *Key methods in geography*, eds. G. Valentine and N. Clifford, 37–53. London: Sage.

Hay, I., and P. Foley. 1998. Ethics, geography and responsible citizenship. *Journal of Geography in Higher Education* 22 (2): 169–83.

Hertz, R. 1996. Introduction: Ethics, reflexivity and voice. *Qualitative Sociology* 19 (1): 3–9.

Illinois Institute of Technology, Center for the Study of Ethics in the Professions. 2007. Codes of ethics online. http://ethics.iit.edu/codes/coe.html (last accessed 29 August 2007).

Israel, M., and I. Hay. 2006. *Research ethics for social scientists: Between ethical conduct and regulatory compliance.* London: Sage.

MacDonald, C. 2002. A guide to moral decision making. http://ethicsweb.ca/guide/ (last accessed 29 August 2007).

MacIntyre, A. 1985. Utilitarianism and the presuppositions of cost-benefit analysis: An essay on the relevance of moral philosophy to the theory of bureaucracy. In *Ethics in planning*, ed. M. Wachs. New Brunswick, NJ: Rutgers University, Center for Urban Policy Research.

National Commission for the Protection of Human Subjects of Biomedical and Behavioral Research. 1978. *The Belmont Report: Ethical principles for the protection of human subjects of biomedical and behavioral research.* Washington, DC: United States Government Printing Office.

Oakes, J. M. 2002. Risks and wrongs in social science research: An evaluator's guide to the IRB. *Evaluation Research* 26 (5): 443–79.

Palys, T., and J. Lowman. 2001. Social research with eyes wide shut: The limited confidentiality dilemma. *Canadian Journal of Criminology* 43 (2): 255–67.

Sieber, J. E. 1982. *The ethics of social research*. New York: Springer.

Sieber, J. E., S. Plattner, and P. Rubin. 2002. How (not) to regulate social and behavioral research. *Professional Ethics Report* 15 (2): 1–3.

Tri-Council (Canada). 1998. *Tri-Council policy statement: Ethical conduct for research involving humans*. Medical Research Council of Canada, Natural Sciences and Engineering Research Council of Canada and Social Sciences and Humanities Research Council of Canada.

———. 2003. *Policy statement: Ethical conduct for research involving humans*. Ottawa: Public Works and Government Services.

Urban and Regional Information Systems Association (URISA). 2003. A GIS code of ethics. http://www.urisa.org/about/ethics (last accessed 29 August 2007).

CHAPTER 13

Academic Publishing

Stanley D. Brunn

Aspiring professionals in the academy as well as those entering the public and private sectors are usually expected to be able to initiate or collaborate with others on research projects that advance a field or subfield. While a good first step in reporting findings and getting feedback is to present to appropriate audiences at workshops and conferences, the major method of communication with colleagues is through publication in disciplinary and interdisciplinary journals (Kenzer 2000), thereby creating a permanent record of the research and reaching wider audiences. Although academic professionals are trained to perfect their writing and presentation skills in graduate school, a logical and important advance for early career professors in the social and environmental sciences is to publish their research (de Souza 1988; Hanson 1988; Kitchin and Fuller 2003, 2005). This goal, however, presents challenges beyond just writing, for both neophytes and experienced academics. In this chapter I offer suggestions and advice about publishing your research.

While the focus of this chapter is on those taking up college and university faculty positions in the U.S. and Canada, much of my advice also applies to those entering the public or private sectors (Boice 2000). Some of the points I will raise apply to academics entering the job market for the first time because many institutions want to see evidence (sometimes multiple examples) of a candidate's pre-Ph.D. publications and presentations. I want to unravel some of the myths, mysteries, and misconceptions about publishing, including "publish and/or perish" (see Kitchin and Fuller 2003; Linton and Embrechts 2007). I will address preparation of manuscripts for peer-reviewed journals, not research monographs, books or book chapters, or editing books. The review processes for books and chapters are often quite different from those of journals. There are, however, useful resources for authors aiming to write a book such as the advice offered in Dedi Felman's column (2006) in *The Chronicle of Higher Education,* a short and readable introduction to what editors look for and some ways in which a book is different from a dissertation.

WHERE DO IDEAS COME FROM?

Identifying a legitimate and important research problem is essential if you hope to get your ideas in print. It is not only the idea or topic being investigated that is important, however, but how you present it. A very legitimate topic could be identified, but if it is poorly articulated and presented, the reviewers and editors may not support publication.

Ideas for early manuscripts commonly come from a dissertation or a thesis, which can be mined for one or two different manuscripts that focus on the major findings and/or on some innovative methodology or model. Additionally, by the time graduate students complete their degrees, they have usually written a half-dozen research papers, each of which might be a possible publication. As an academic progresses, ideas also stem from interacting with colleagues, including those in related fields, conducting seminars, listening to conference presentations, reading books, articles, book reviews (very good sources), presidential addresses, conversations on field trips, and exchanges on listservs. It is useful to keep a list of potential research topics in a folder or notebook (printed or electronic) and to update it with ideas, references, potential funding sources, maps, conversations, valuable web sites, informative footnotes, and listserv discussions.

WHY PUBLISH?

This question is raised in graduate school and by those seeking first jobs. I can think of three major reasons for publishing. First, job descriptions and performance evaluations usually include publication as an expectation. No one should assume a position without first knowing what the expectations are for research and publication. Usually, they are greater in major research universities with doctoral programs than in institutions where heavier teaching and advising are the norm. While the amount and type (articles, chapters, or reports) will vary with the job and title, publishing is expected for retention and promotion; funding agencies that support research also look to see if results of previous awards have been disseminated via publication. The research expectations for temporary and part-time faculty are less than those for full-time positions, but since people in such situations may be aspiring to full-time and permanent academic appointments, they need to consider whether and how they will make efforts to publish.

A second reason for publishing is the enjoyment that comes from working with ideas, writing, and sharing results. Many academics get excited about starting a new project, even if completing it requires much time and money (their own or from other sources), and then preparing presentations and manuscripts for journals. I have observed junior and senior geographers wax with enthusiasm and eloquence about their research projects on topics such as climatic change models, gender and work, and Internet mapping. Many are stimulating teachers who are not simply driven by "requirements" for promotion and tenure as they undertake multiple projects simultaneously and publish several articles yearly.

Third, publishing is widely required for professional advancement (see also Susan Roberts's chapter on "Succeeding at Tenure and Beyond"). In most institutions, retention and reappointment are contingent on maintaining and demonstrating an active research program. The degree of activity will vary with the university's mission. Promotion to associate professor requires an accounting of what one has accomplished as an assistant professor. That record should reflect intellectual growth beyond the dissertation research. Some colleges and universities may require only one or two published items, but others may expect two or three papers in major refereed journals each year. Additional research-related materials considered for promotion will be book chapters or books (edited or co-edited), research proposals (both funded and rejected), reports for governmental and private clients, book reviews, and conference presentations. While the formula for promotion to full professor will be similar to that for an aspiring associate professor, and also vary by institution, the expectation is that the research record during the time in associate rank will have shown sustained development; research universities will also expect that some national and international stature has been achieved for a full professorship. Publications are a major indicator of such recognition.

WHEN AND WHAT TO PUBLISH

Because many current job advertisements specify that applicants have a record of publication prior to earning the doctorate, it is important to begin thinking about publishing while in graduate school. This requirement is often the threshold even for being seriously considered, with applicants lacking publications often placed in a different/lower category. For aspiring academics, a first manuscript might come from a seminar paper or thesis that is presented at a regional conference and submitted to a regional journal. As the graduate student advances, it is important to think about building on this experience, to present at national conferences, and to consider submitting an article to a journal that has wider recognition or circulation.

The dissertation is the obvious place to look for early and substantial publications that establish your status as a scholar because it is a sustained piece of work that has survived the scrutiny of an advisory committee, which validated its worth. The committee's advice is valuable in recommending not only what to publish (major findings, the methodology, the techniques, etc.) but also where to submit.

WHERE TO PUBLISH?

Many authors consider this question when they embark on a project. Some will submit their manuscripts to refereed journals, others for non-refereed conference proceedings, though these are often less valued, especially in institutions that offer doctoral degrees. Since professional journals are "information gateways" for future scholarship, the outlet becomes very important. The recipes vary by subfield: human geographers, for example, commonly write sole-authored articles and may write for one of the many specialized journals in historical, cultural, political, or economic geography. Those in physical geography and technical specialties, whose research may be done in teams, often write coauthored articles and send their articles to interdisciplinary journals. Bourne (2000) offers useful reflections on experiences of publishing in human geography, and Butler (2000) in physical geography. Another question to consider is whether findings are best shared with one's own discipline, with scholars in another field, or with those whose major language is not English (Turner 1988). Some authors check the impact ranking of a journal and submit manuscripts first to those that have high ratings based on the extent to which the articles published in them have been cited by other scholars as identified in citation indices such as the ISI Web of Knowledge.

Young professionals often aspire to publish their findings in highly ranked and peer-reviewed international journals, advice that probably came from their advisor and members of their dissertation advisory committees. Publishing in these journals gives the author wide visibility, but I maintain that very few articles from dissertations appear in them, perhaps because they were not submitted, they were rejected, or the authors submitted them elsewhere. While flagship and high-impact journals can be important in a research career, many lesser-ranked journals regularly publish quality and innovative research, commentaries, and book reviews. A check of the references in articles in major journals will reveal many citations from the more specialized journals that have proliferated in recent years, as well as from lesser-known and lower-impact thematic, regional and interdisciplinary journals, and from books. The approach that a journal favors is also a key criterion. Does it feature articles that are strongly theoretical, ones that emphasize empirical studies, or those that offer a balance? Does it focus on particular paradigms or methodologies?

It also pays to be thoroughly familiar with a journal's orientations prior to submission. Scholars should not be deterred from seeking to publish their research by thinking that only those from major research departments will be included in journals. Good, solid, and frequently cited publications come from authors in doctoral, master's, and baccalaureate programs and in large and small departments and universities (Brunn 1995), and there are many prolific professors in baccalaureate programs in which the teaching and advising responsibilities are heavy.

AUTHORSHIP

Questions often arise about who and how to list as authors of an article. There are several issues. First, if the manuscript is based on your dissertation or seminar paper, you legitimately should be sole author. This point may be ticklish because some advisors want to be counted as coauthors; they may expect to be recognized because the dissertation topic or seminar paper idea originated in their classes, or the research was supported under a grant that they had been awarded. Some young scholars who want and/or need a pre-Ph.D. or an early post-Ph.D. publication welcome a dual authorship arrangement with their advisor, knowing that this may be a diplomatic approach. Other advisors are very comfortable with their advisees preparing potential manuscripts as sole authors and experiencing the process themselves. These advisors are content to be listed in the acknowledgments section.

Second, it is prudent early to agree on the "order" of multiple authors. I know two-member author teams who alternately switch the order of authors. In multiple-authored papers, a common guideline is for the person who contributed the most time and effort, or who obtained the grant, to be listed as the lead author. When contributions are equal, the order is usually alphabetical. As noted earlier, multiple authorship is more common in the physical (and medical and behavioral) sciences than in the social sciences. In the humanities, single-authored publications, especially books, are highly important in promotion decisions. An early career professional should check with a potential employer about what will "count" when he or she is evaluated for purposes of salary increases and promotion.

REWARDS

There are many joys awaiting those who publish. These rewards come from seeing long hard work being completed, from a sense of individual accomplishment in completing a task, and from insights gained in collaborative explorations. Many junior scholars and career professionals find that having something published, either in a major or minor journal, is both professionally and personally rewarding (all authors remember their first acceptance and rejection letters). When a manuscript is published, it means your peers like your work and want to see it disseminated. Probably next in gratification to having an article, research note, or book review published is seeing one's own work cited by prominent or unknown professionals in one's discipline and beyond.

Publishing several articles early in one's career and on related topics will provide name recognition to editors, peers, program directors and reviewers at funding agencies, and prospective employers. The recognition that comes from publishing significant, quality, and cutting-edge research may result in some unforeseen professional benefits and surprises, including being sought as a candidate for a job, and invitations to participate in disciplinary and interdisciplinary conferences and workshops, or to submit a paper for publication in a thematic journal issue. Publications should result in salary increases.

EDITORS AND THE MANUSCRIPT REVIEW PROCESS

Editors are crucial individuals in the early years of an academic career. They come in various shapes and sizes, just like members of the organizations they serve and the universities they represent. Some are relatively junior in ranking, others have senior standing; some are appointed by publishing houses and universities, others are selected by professional societies (the Council of the Association of American Geographers, for example, advertises for and selects the editors of the *Annals of the Association of American Geographers* and *The Professional Geographer*). All editors have favorite topics and personalities that range from positive and engaging to being perhaps haughty and unpleasant. They serve as disciplinary gatekeepers in that they are responsible for what others see about a discipline's scientific and theoretical advances.

Successful editors like to work with ideas and with authors; they also tend to be independent and seek to be fair. They recognize their decisions may have critical implications for a scholar's career. The editor of a refereed journal (not all journals have a review policy that is mandatory and transparent) selects reviewers and makes decisions based on the advice of a number of individuals, including editorial board members and others selected because of their familiarity with the subject matter and because they have a reputation for providing prompt and constructive reviews and no apparent conflict of interests with the author.

Commonly, a manuscript is sent to three reviewers (though some journals seek more) who are asked to respond within a specific time and given a set of criteria to consider in their evaluations. Editors read manuscripts independently and study reviews and the reviewers' recommendations (Brunn 1988). While there is often a consensus among reviewers on the manuscript's merits and shortcomings, when discrepancies occur, the editor evaluates what weight to give to the different opinions. The editor's decision commonly is in one of three categories: "accept," "revise and resubmit" (R and R), and "reject." Some offer the alternative "accept with minor revisions." Indeed, very few manuscripts are given a strong "accept" initially. The R and R is "positive," signaling that the paper has merit, but requires additional work before possible publication.

When manuscripts are rejected it is usually because reviewers have judged them to be flawed in writing, methodology, analysis, and/or presentation. Those with potential (R and R) may require a more comprehensive literature base, a stronger theoretical section, clarification of the methodology, additional analyses, and/or tighter writing. The editor shares the reviewers' comments with the author without revealing the reviewers' identities; he or she usually writes a detailed letter summarizing the key points to address in revision.

A revised manuscript should be carefully thought out and prepared. When it is submitted, it should be accompanied by a detailed (point by point) cover letter that addresses specific and general points raised by reviewers and the editor. The revision would likely be sent to at least one previous reviewer and perhaps two new ones. When some editors send the revised paper to reviewers who have already read the first submission, they enclose the original set of reviews to facilitate assessment of whether the criticisms and suggestions have been adequately addressed. It is not unusual for manuscripts to be subjected to a third round of revisions.

The length of time for a manuscript to be published following initial submission depends upon when reviewers respond, the time taken to prepare the revision, the time required for subsequent reviews, and the backlog of accepted manuscripts. Once a manuscript is accepted, the paper might see the light of day within six months but the time to publication may be as long as eighteen months. Some editors may publish the paper earlier, especially if they are looking for a manuscript that fits the exact page allotment for an issue or if it provides some balance to the readership. Usually editors submit materials to a printer several months before a specific number appears. Some journals now offer online preprints of accepted articles in order to speed the circulation of research.

The process of manuscript submission, review, and responding to an editor's suggestions for revision is illustrated in Activity 13.1 on this book's web site—this activity will teach you some of the ways to work effectively with journal editors.

A CHECKLIST OF HINTS FOR AUTHORS

Editors can readily provide a specific list of hints regarding manuscript preparation, electronic submission, and revisions. It is important to visit the journal's web site for specific information prior to submission of a manuscript and to study the instructions for authors. Journal editors also often offer advice panels at professional conferences. To supplement such hints, I offer the following additional recommendations:

1. Plan a pre-tenured publishing schedule that includes a mix of grant proposals together with articles, chapters, and research notes with tentative deadlines and tentative journals.
2. Ensure that the manuscript submitted is original and that it is your own work. State in the cover letter to the editor that this or a similar manuscript has not been submitted elsewhere and will not be sent elsewhere until you hear from the editor.
3. Provide the editor with (brief) information regarding the genesis of the manuscript, your academic background and research interests, whether the manuscript is based on a thesis or dissertation, field

research (done when and where), the source of any funding, and anything else you think might be useful for the editor to know.

4. When submitting a revision, be sure the cover letter contains detailed responses to points raised by the reviewers and the editor. You need not agree with them (state why), but you need to explain succinctly and clearly why you do not.

5. Study carefully the journal's instructions for authors, including details related to preparation of figures, tables, any special word-processing requirements, and whether to submit on paper, in electronic form, or both. Editors may return manuscripts immediately if they fail to conform to the journal's requirements.

6. Include specific sections that may be called for in the instructions, for example, a short title, detailed abstract, key words, theoretical or conceptual statement, and methodology. Consult previous issues for major headings and subheadings. Include an acknowledgment that lists funding sources, support from your university, cartography lab, advisors, and others.

7. Check that all bibliographic entries conform to the journal style and that all appear in the text and vice versa. This can be quite a frustrating and time-consuming process because journals vary in their styles.

8. Be sure the manuscript is well written, is grammatically correct, and has correct spellings (including accents) of all words from other languages. Be sure accurate captions accompany photos, maps, and graphics.

9. Identify a mentor (preferably a senior professor with experience) who can/will assist you in your professional career with advice on grant submissions, journal outlets, teaching/research/service balance, and time commitments.

10. Ask your advisors and impartial colleagues to read your manuscript before it is submitted, including people not familiar with the subject matter. Let them know where you wish to send the manuscript. Also feel comfortable in working with professionals in your university's writing center about manuscript preparation and revision. Their assistance may make the difference between rejection and an R and R decision.

11. Work with the editor on all phases of the manuscript review and revision process. If your manuscript is rejected, feel comfortable knowing why.

12. Be sure that the manuscript is of the highest quality, not marginal in scholarship in any way—analysis, methodology, techniques, graphics, and so on. Hold to the highest of ethical and professional standards at all times (Brunn 1989).

13. Avoid manuscript cloning and submitting multiple niche manuscripts; better to have two solid contributions than four niche papers (Brunn 1998; Graf 2004).

14. Do not plagiarize, that is, include small or large sections of text, maps, or photos without appropriate, correct, and complete citation, or self-plagiarize, that is, lift (include) large sections verbatim of another work you have published.
15. Do not include a supportive letter from your advisor or possible reviewers.
16. Make moderate use of self-citation, especially when necessary to avoid self-plagiarism. Also do not list "forthcoming" items in the list of references; you can inform the editor of relevant forthcoming manuscripts in your cover letter.
17. Do not bribe or attempt to influence the editor or badger the editor with repeated e-mails or phone calls about the review process. Avoid getting your advisor or department chair to intercede on your behalf. Delays are usually due to tardy reviewers.
18. Do not be obsessed with publishing by such approaches as seeking to publish quickly, confusing quantity with quality, and using journal rankings only to build a research career.

SOME FINAL "WORDS OF WISDOM"

Finally, I include some additional advice from conversations with young professionals who have successful publishing careers:

1. For non-tenured, adjunct, and part-time faculty in colleges and universities it is probably unwise to write textbooks during one's first few years as a professional. Research monographs and dissertations published by major commercial publishing houses or university presses may "count."
2. Editing books and authoring lab manuals are unwise projects for assistant professors. Wait until you have advanced in rank.
3. Almost (probably) all authors, junior and senior, well cited and with international stature, have had manuscripts rejected. Those who say otherwise are being less than candid. Success always comes with some failures. If a manuscript is rejected, seek out mentors for your next steps.
4. Publishing is a lifelong learning process. One should strive to become a polished writer and professional with each manuscript written. To look at academic publishing in any other way is not very instructive.
5. There are many quality outlets. These include flagship and minor journals, those in English and other languages, and those with interdisciplinary audiences.
6. For every good manuscript, there is an appropriate journal; it may require some time to identify the "right" one. Patience and persistence are keys to success in any publishing career.
7. A professional career is seldom based on a publish-or-perish model, but rather on all facets of a job description.

8. A mix of publications is desirable in one's career: some single-authored, others multiple-authored, grants, conference presentations, book reviews, and edited books.
9. Those who produce high-quality work on a sustained basis will be recognized for their efforts. Sloppy and shoddy research (if it does get published) and suspicions that data are falsified or materials plagiarized may cause you to suffer professional consequences (loss of job, graduate advisees, research dollars, and respect).

The meaningful professional life has constant growth and benefits, including working with ideas and with people. That life includes identifying interesting and timely research projects and communicating those results in oral or written form. A certain enjoyment comes from preparing a manuscript for one's peers. The manuscript process requires patience, persistence, and flexibility on the part of authors before their ideas will see the light of day. The end result should be a product that you, as an early career professional, are pleased with and one that you envisage as a contribution (minor or major) to a given field. While the benefits of publishing early and often are immediate and apparent in job placement and security, over the long term a sustained record of quality research will lead to some unexpected and challenging opportunities for intellectual and personal growth.

References

Boice, R. 2000. *Advice for new faculty members.* Boston: Allyn & Bacon.

Bourne, L. S. 2000. On writing and publishing in human geography: Some personal reflections. In *On becoming a professional geographer*, ed. M. S. Kenzer, 100–12. Caldwell, NJ: Blackburn Press.

Brunn, S. D. 1988. The manuscript review process and advice to prospective authors. *The Professional Geographer* 40 (1): 8–14.

———. 1989. Editorial: Ethics in word and deed. *Annals of the Association of American Geographers* 79: iii–iv.

———. 1995. Research performance based on *Annals* manuscripts, 1987–92. *The Professional Geographer* 47: 204–15.

———. 1998. Quality research performance: Issues of splitting, cloning, citing, significance and judging. *The Professional Geographer* 48: 103–5.

Butler, D. R. 2000. Conducting research and writing an article in physical geography. In *On becoming a professional geographer*, ed. M. S. Kenzer, 88–99. Caldwell, NJ: Blackburn Press.

de Souza, A. 1988. Writing matters. *The Professional Geographer* 40:1–3.

Felman, D. 2006. What are book editors looking for? *The Chronicle of Higher Education,* 21 July. http://chronicle.com/weekly/v52/i46/46c00101.htm (last accessed 21 July 2006).

Graf, W. L. 2004. Fakery in the publications game. In *Presidential musings from the meridian*, eds. M. D. Nellis, J. Monk, and S. L. Cutter, 238–41. Morgantown, VA: West Virginia University Press.

Hanson, S. 1988. Soaring. *The Professional Geographer* 40:4–7.

ISI Web of Knowledge. http://portal.isiknowledge.com.

Kenzer, M. S. ed. 2000. *On becoming a professional geographer.* Caldwell, NJ: Blackburn Press.

Kitchin, R., and D. Fuller. 2003. Making the "black box" transparent: Publishing and presenting geographic knowledge. *Area* 31:313–15.

———. 2005. *The academic's guide to publishing.* London: Sage.

Linton, J., and M. Embrechts. 2007. Unsolicited advice on getting published and the publishing process. *Technovation* 27:91–94.

Turner, B. L., II. 1988. Whether to publish in geography journals. *The Professional Geographer* 40:15–18.

CHAPTER 14

Working Across Disciplinary Boundaries

Craig ZumBrunnen and So-Min Cheong

INTRODUCTION

Growing commitment to interdisciplinary teaching and research is a national phenomenon as funding agencies such as the National Science Foundation (NSF), National Institutes of Health (NIH), universities, and research centers recognize the need for interdisciplinary collaboration in order to tackle complex social and environmental problems. The desire for interdisciplinarity stems from both the theoretical and practical priorities of academia, industry, and society. Unmet societal needs and the practical financial concerns of researchers in the face of declining governmental support for graduate education and research funding, especially for state-funded research universities, are creating conditions favorable for interdisciplinary education, training, and research. Business and public funders are particularly interested in efforts to develop practical solutions to problems that inevitably cut across disciplinary boundaries and geographic scales (Schoenberger 2001). Drawing on the diversity of perspectives and practices each discipline offers, interdisciplinary collaboration promises to provide innovative solutions to complex contemporary problems (Reich and Reich 2006).

An excellent model of interdisciplinary teaching and research training is the Integrative Graduate Education and Research Traineeship (IGERT) program initiated by NSF in 1997. Since then, more than 214 IGERT awards have been made to over 125 award sites. Substantive topics in interdisciplinary work include multifaceted globalization, climate change, human dimensions of global change, natural and human-induced hazards, sustainability, poverty alleviation, and urban growth. Because addressing such themes requires bringing together specialists in human, biophysical, nature–society, and technical realms, and attention to public policy concerns, programs such as IGERT are being established to expand opportunities for graduate students to gain experience working in interdisciplinary teams and research settings.

It is also important to recognize that interdisciplinary partnerships are not unique to the U.S. Such efforts are evident elsewhere and include opportunities for international collaboration in addressing issues of global significance. For example, geographers are exceptionally well represented in new multidisciplinary units in Australian universities (Holmes 2002), while German geographers have been very active in the German Research Foundation's (DFG) IGERT-like GRAKO (Graduiertenkolleg) interdisciplinary research program and have taken part in mutual faculty and graduate student exchanges and collaborations with the University of Washington's urban ecology IGERT program.

In this chapter we focus on emerging paradigms of interdisciplinarity and interdisciplinary praxis and offer suggestions of ways to enhance skills for research, education, communication, and management while working across disciplinary boundaries. We highlight training needs and describe experiential ways of improving them, calling attention to some of the institutional impediments to interdisciplinary work. The chapter is accompanied by a set of practical activities designed to assist faculty and graduate students who wish to be successful in meeting the challenges in interdisciplinary and transdisciplinary team efforts. Our suggestions and the activities draw heavily on our combined forty-plus years of personal experience in crossing disciplinary boundaries as students, teachers, and researchers. Attention to language, differing personal work styles, systems thinking (Senge 1990; Bellinger 1994), experiential problem-based learning (PBL) (Farley et al. 2005; White 2001); teamwork, and formal group-process training exercises (Scholtes, Joiner, and Streibel 2003) are important strategies and activities for improving the skills needed for intellectually stimulating, rewarding, and successful interdisciplinary work.

IMPORTANT DEFINITIONS

Collaborations that involve working across disciplinary boundaries are often described as multidisciplinary, interdisciplinary, or transdisciplinary. What do these terms share, how do they differ, and what research and learning strategies and approaches are most effective for such efforts to be creative and successful? All three terms refer to efforts by natural and social scientists and scholars from the humanities and professional fields to address a myriad of complex issues and problems in which geographers and other scholars are, or need to become, engaged in order to make useful theoretical and practical contributions to their resolution.

At the most basic level, "multidisciplinary" approaches are contexts in which two or more researchers from two or more disciplines work collaboratively on a problem or set of problems without significantly modifying their respective disciplinary approaches or developing truly synthetic frameworks. In contrast, the successful applications of "interdisciplinary" approaches are those in which researchers from two or more disciplinary backgrounds come together to use innovative conceptual frameworks to modify and synthesize their respective disciplines in order to deal effectively with a research problem (Alberti et al. 2003). "Transdisciplinary" approaches evolve in meaning and actively engage nonacademic practitioners in collaboration with academic researchers to identify, research, and develop solutions to real-world problems (Tress, Tress, and Fry 2003). Such approaches employ a combination of appropriate theories and methodologies to solve problems with the problem being addressed determining the tools and disciplinary knowledge required to develop appropriate solutions (Graybill et al. 2006).

KEY ASPECTS OF CROSSING DISCIPLINARY BOUNDARIES

Interdisciplinary Research

Since the 1980s, and with the end of the Cold War, research institutions have been buffeted by a cacophony of political, social, and economic changes that have shifted dwindling research funds away from "basic" research to more "strategic" research based on societal needs (National Research Council 1994). As scholars and researchers, we are now frequently called upon and expected to contribute to ongoing and new debates on public policy. This has created an endless need for individual and disciplinary reflection and generated debate surrounding questions of advocacy versus research objectivity (*e.g.*, see, Sonnert 2002; Campbell 2005).

Undertaking interdisciplinary research can be a lengthy process. It takes time to learn about other disciplines and bodies of research and to come to a consensus about ways to collaborate and implement research goals. Herbert Simon considers interdisciplinary collaboration a serious business and estimates that it takes at least one year to "pick up" another person's discipline (quoted in Lattuca 2001). It also takes time to develop a shared cognitive space or paradigm that allows participants to think about the world in similar ways (Rossini and Porter 1984). (We suggest an approach to developing such thinking through undertaking a cognitive mapping exercise; see Activity 14.3.) This process is crucial, however, as it fosters shared standards and agreement on methods. For example, in the case of interdisciplinary human–environment research, the complexity of human and environment interactions requires a multimethod framework where computational modeling, statistics, qualitative analysis, and geographic information systems, among other methods, come together in order to achieve as comprehensive an analysis as possible. Cutting-edge methodologies for integrating human and natural systems are currently being tested to see if shared standards and methods are within reach (Alberti et al. 2003).

Management of Interdisciplinary Research

Monk, Manning, and Denman (2003) identify several key components of effective teamwork for interdisciplinary projects. Identifying the partners through networking among academic and nonacademic individuals and setting a shared agenda on goals and objectives are important beginnings for a productive collaboration. Seminars and discussions, preferably face-to-face, enhance mutual understandings and serve to reveal and acknowledge differences in approaches and disciplinary knowledge. It is also important to rotate meeting sites. Developing a division of intellectual labor based on individual or institutional strength is the next step, dovetailed by the assignment of management responsibilities including budget specifications and distribution of funding. Through these processes, collegiality and trust in partnerships are built and maintained (a theme that is addressed further in Chapter 3 "Developing Collegial Relationships in a Department and a Discipline" by Duane Nellis and Susan Roberts). Writing and monitoring progress toward goal attainment are important elements as the collaboration progresses.

A critical component of teamwork is spending time together in work, social, and recreational contexts. Regularly scheduled gatherings and dinners are an effective and fun way to promote the building of cohesive and successful interdisciplinary learning and research teams. Seeking the assistance of a professional team coach or facilitator early in the collaboration to lead daylong or multiday skills and team-building workshops may also be helpful, especially if researchers are involved in a large and lengthy research project. Although such team-building activities may take time, they enhance and hasten the creation of interdisciplinary knowledge, learning, and research (in Activities 14.1 and 14.2 we offer some strategies for developing and sustaining team relationships).

Interdisciplinary Training

Increasing interest in interdisciplinary work is also influencing graduate education. Scholars have called for new directions in graduate education and training in order to prepare researchers and educators for the twenty-first century (Committee on Science, Engineering, and Public Policy 1995; Nerad and Cerny 1999; Nyquist, Manning, and Wulffe 1999; Nyquist 2000; Golde and Dore 2001; Pallas 2001; Lélé and Norgaard 2004). Despite their diverse perspectives, many of these recent studies emphasize the importance of

(1) increasing the versatility and career options of Ph.D. candidates; (2) training in teamwork and managerial skills, including those commonly required in business, industry, and private and nongovernmental sectors; (3) participating in internships; (4) providing more career assistance; and, most importantly, (5) encouraging interdisciplinary (and often international) work.

Other important recommendations include the call for increasing exposure to technology, extensive review of program requirements, program assessments, and clarification of program expectations for graduate students. Further recommendations include better mentoring for students, new reward and incentive structures for faculty engaged in interdisciplinary work, formal course work on values and ethics involved in research and teaching, and the incorporation of formal training and understanding of the global economy and environment. If these programmatic improvements were to be implemented across graduate education, the next generation of scholars would be better prepared to engage interdisciplinary projects.

Curriculum Design

In order to design effective interdisciplinary training and curricula, several procedures are necessary. The first step is to create introductory courses on both substantive and methodological topics that will make it possible for students to acquire basic skills (Goldstein et al. 2004). Second, a core set of courses that provide a broad background coupled with specialized courses drawn from different disciplines should be established (Semerjian et al. 2004). Third, suites of actively team-taught, interdisciplinary, issue or problem-based thematic courses need to be developed (discussed in more detail in the next section). The objective of such courses is to encourage students to develop expertise on an issue by "thinking across disciplinary lines" (Caviglia-Harris and Hatley 2004, 396). In addition, this allows for interdisciplinary feedback as students see similarities and differences among disciplines and begin to recognize how different disciplines approach the same problem (Manathunga, Lant, and Mellick 2006). Finally, students must have easy access to interdisciplinary curricula. This requires good advertisement and administrative support to enable students to enroll in such programs and courses (Semerjian et al. 2004).

Aside from course development, interdisciplinary exchange fellowships, which encourage graduate students to study in another discipline for a year, are useful (Manathunga, Lant, and Mellick 2006). The inclusion of guest researchers and speakers and active discussion of epistemological, vocabulary, value, scholarly ethics, and procedural differences and similarities across disciplines in and out of classrooms should also be emphasized (Shearer 2007). Implementing fieldwork, internships, and community work through service learning courses, as well as post-course work, also support the goals of interdisciplinary and multidisciplinary work (Semerjian et al. 2004).

Teaching Interdisciplinary Courses

The objectives of interdisciplinary courses are to introduce students to complex multidimensional issues, to develop a holistic understanding of problems/concepts/ideas, and to create projects that address real problems. The application of interdisciplinary work is crucial in this regard (Semerjian et al. 2004). Systems thinking and PBL are strongly encouraged as ways to approach complex, real-world problems; to motivate students to identify and research important concepts and principles; and to acquire the necessary skills, theories, and data to develop solutions (Senge 1990; Bellinger 1994; Duch, Groh, and Allen 2001). Although geography's holistic and place-based approach is an ideal candidate for the proliferation of PBL techniques (Chappell 2006), students are rarely exposed to PBL in traditional geography courses (Pawson 2006). We therefore offer an example of PBL in Activity 14.4.

Shearer (2007) provides useful insights for developing interdisciplinary courses using the example of a cross-disciplinary, team-taught, science ethics class. The steps required in course planning include decision-making regarding content, assessment, and practical arrangements. It is important to agree on course goals, objectives, and topics and to determine who will teach each section. The planning will involve holding numerous meetings to ensure that each approach is allocated an appropriate amount of time, that topics are complementary, and that teaching does not involve unnecessary duplications. Discussion also needs to address how each participating discipline assesses knowledge and to reach consensus on assessment.

Another tool for teaching interdisciplinary courses is to conduct "interdisciplinary discussions" seminars where professors from multiple fields are continuously present and students are encouraged to discuss from each of these perspectives. Such seminars allow students to see differences and similarities between approaches (Caviglia-Harris and Hatley 2004). One particularly effective approach to interdisciplinary teaching is to use hands-on activities in which students bring systems-thinking theory and practice to bear (Senge 1990; Bellinger 1994; Goldstein et al. 2004; and see Activities 14.3 and 14.4).

Student Challenges

Students in interdisciplinary programs often struggle with their "identity" since they may not be part of a traditional departmental degree program. Furthermore, graduate students in interdisciplinary programs are often not eligible for traditional fellowship and traineeship opportunities. In some cases, graduate students teach two courses simultaneously both to satisfy the interdisciplinary program requirements and to stay funded by their home departments (Graybill et al. 2006). Such students may also have difficulties identifying and recruiting thesis committee members because, although they are intellectual collaborators, faculty members may not necessarily get credit in their home department for interdisciplinary students.

INTERDISCIPLINARY PROGRAMS

What Makes a Good Interdisciplinary Program?

Financial and institutional supports are critical for initiating and sustaining interdisciplinary teaching programs and conducting interdisciplinary research and outreach projects. Equality in participation, contributions of financial resources and their allocation, and equitable decision-making across participating institutions and programs are important (Monk, Manning, and Denman 2003). If there is administrative resistance and indifference (from presidents, deans, and department chairs) programs will have difficulties (Harris et al. 2003).

The ability to share is also an important feature of good interdisciplinary endeavors. As mentioned above, objectives, goals, knowledge, and experience must be communicated through frequent information sharing (Harris et al. 2003). Sharing "turf," building trust, and avoiding disciplinary ethnocentrism also help to prevent protectionist barriers between participants (Harris et al. 2003; Monk, Manning, and Denman 2003). Hence, the ability to communicate is a must. Again, it is important to make sure to spend time at the beginning of any collaboration developing shared understandings of concepts and ideas that promote a shared language (Bracken and Oughton 2006).

The core elements of successful interdisciplinary programs and projects are leadership, complementary missions, faculty development, and feedback (Harris et al. 2003). They require leaders who encourage participatory governance, can advocate across disciplines, share power and influence, are enthusiastic, and are able to cultivate others' enthusiasm. Effective collaborations are able to enhance the missions of other institutions, be they within the university or the community. For example, an environmental science program enabling students to participate in activities that helped improve the access and delivery of primary care in local communities not only led to greater support for the program but also resulted in positive learning experiences for the students (Semerjian et al. 2004). Efforts to reduce turf issues can also be made by fostering project teams that focus on multidisciplinary case studies and include a variety of participants such as students, field professors, and campus-based faculties. Getting feedback continually from program participants is also important in order to assess the effectiveness of programs (Harris et al. 2003; Semerjian et al. 2004).

How Are Good Interdisciplinary Degree Programs and Research Projects Developed?

Several good leads on the development of interdisciplinary programs and projects are available. For example, Monk, Manning, and Denman (2003) describe five phases in the process of developing sustained interdisciplinary research, faculty development, and outreach effort on the topic of gender and women's health along the U.S.–Mexico border. These included networking for partners, exploration to develop a common agenda and vocabulary, planning for funding and personnel resources, building for additional participation, and consolidation/transformation of the project. Graybill et al. (2006) identify three major stages—naissance ("where is my home?"), navigation ("what do I prioritize?"), and maturation ("how do I integrate and represent my scholarship?")—in an assessment of graduate student participation in the University of Washington's urban ecology IGERT program. They indicate that having a good experience in an interdisciplinary program begins by getting to know both the home department and the program through coursework, degree requirements, and interpersonal communication. In the navigation period, graduate students learn to balance and reconcile dual intellectual and institutional requirements, and to integrate the two systems. At the maturation stage, students' theoretical and practical knowledge culminate in their dissertations and in publications in suitable interdisciplinary journals. For those interested in research on the challenges of institutionalizing interdisciplinarity on university campuses and the future of interdisciplinary in general, Julie Thompson Klein's works (*e.g.*, 1996, 1999, 2004, 2005) and the entire May 2004 issue of *Futures* would be useful.

Strategies for Internal and External Interdisciplinary Support

Several strategies for generating and sustaining internal and external support for interdisciplinary programs and projects have been identified:

- Institutional assessments of individual accomplishments need to recognize and reward collaborative work (Monk, Manning, and Denman 2003).
- Departments need to recognize interdisciplinary teaching contributions— often these courses are not considered "core" courses and become categorized as additional teaching service, which creates time pressures for faculty (Shearer 2007).
- Interdisciplinary cultures that support the goals and objectives of such work also need to be developed within participating units (Schoenberg 2001).
- Networking should include individuals external to the academy. By identifying program directors and personnel from community and funding organizations, collaborators are better able to generate external support for interdisciplinary programs and projects. For example, program officers from funding institutions should be invited to seminars and conferences in the exploratory and planning stages to help spark their interest in participation and support (Monk, Manning, and Denman 2003).

Institutional Challenges Faced by Interdisciplinary Programs

While interdisciplinary work does not require the existence of formal unit programs, academic and other research institutions are increasingly trying to develop them in order to promote and facilitate new endeavors. Common challenges according to DuBrow and Harris (2006) include the following:

- *Faculty hiring, promotion, tenure, merit increase, and retention issues:*

  1. Concerns about the ability to be promoted, especially for junior faculty, make joint faculty appointments difficult. Because their home departments often do not reward such efforts, senior faculty may not be motivated to participate in interdisciplinary programs.
  2. Participation in such programs is not uniformly acknowledged across departments or rewarded when promotion and tenure decisions are made.
  3. When departments provide merit raises above the average, interdisciplinary programs rarely have the funds to match these raises.
  4. Department chairs and deans are not rewarded for encouraging participation in interdisciplinary programs by their faculty.
  5. Faculty beginning work in an interdisciplinary unit, or with a joint appointment in a disciplinary and interdisciplinary unit, often face a number of critical questions that may be controversial, ignored, or delayed, including the division of time, access to resources, how and by whom they will be evaluated, and even where they will have space.

(For discussion of related issues, see Chapter 5 "Succeeding at Tenure and Beyond" by Susan Roberts in this book.)

- *Resource allocation:*

  1. Interdisciplinary programs often do not receive the standard indirect costs from research projects that result directly from their initiatives; advance "deals" need to be made between departments and schools to share these funds.
  2. There are few ongoing sources of support for interdisciplinary programs. Even those programs that are well established often lack mechanisms to raise funds within current institutional structures. This situation inhibits the creation of "intellectual space" for faculty and students to experiment and to take risks with new academic directions.
  3. Somewhat counterintuitive is the common experience that it is almost impossible to eliminate interdisciplinary programs even if they cease to be effective and are not meeting their research and/or teaching goals.

- *Program advocacy:*

  1. Development activities such as fund-raising for interdisciplinary programs are often viewed to be in competition with departments and schools/colleges involved with the programs.
  2. There are rarely effective advocacy procedures or clear reporting lines for interdisciplinary programs within most universities' organizational structures.

CHOOSING TO CROSS DISCIPLINARY BOUNDARIES IN YOUR CAREER

Making Career Choices and Assessing Career Impacts

Table 14.1 summarizes how a group of nearly 900 researchers ranging from graduate students to professors and principal investigators from five programs including a Human Dimensions of Global Change Center, two IGERT programs, a National Synthesis Center, and a Science and Technology Center viewed the career effects of interdisciplinarity. Graduate students were most likely to participate in interdisciplinary collaborations (62 percent reported participating in at least one interdisciplinary collaboration compared to 49 percent of professors). Seventy-two percent of all respondents and 67 of graduate students reported positive effects of participation in interdisciplinary projects. At the same time, 16 percent of graduate students as opposed to only 10 percent of professors who reported negative impacts (10 percent of postdoctoral fellows also reported negative impacts) of their interdisciplinary participation. Overall, graduate students considered their participation as far more intellectually rewarding and practical in providing solutions to the problems of society than professionally risky.

TABLE 14.1 Views on career effects of interdisciplinary research.

Distribution by rank	G	NTT	PD	AsP	AP	P	PIs	Total
Number surveyed	160	245	84	73	82	232	12	888
Total responses	99	155	59	47	53	147	11	571
Positive	67	104	42	34	43	109	11	413
	68%	67%	71%	72%	81%	74%	100%	72%
Neutral	16	43	11	12	8	23	0	114
	16%	28%	19%	26%	15%	16%		20%
Negative	16	8	6	1	2	15	0	44
	16%	5%	10%	2%	4%	10%		8%

G, graduate students; NTT, nontenure track; PD, postdoctoral fellow; AsP, assistant professor; AP, associate professor; P, professor; PIs, principal investigator. Table modified to include cell percentages.

Source: Rhoten and Parker 2004.

Summary of the Benefits and Challenges of Working Across Disciplinary Boundaries

Effective collaboration in interdisciplinary research and education prevents narrow specialization and provides contextual knowledge that better addresses complex, real-life issues (Manathunga, Lant, and Mellick 2006). Some argue that interdisciplinary learning leads to more flexible thinking, greater awareness of the strengths and weaknesses of a discipline, improved cognitive skills and higher-order thinking, greater awareness and tolerance of ambiguity, and increased sensitivity to bias, in addition to many other positive outcomes (Shearer 2007). For example, Ivanitskaya et al. (2002) provide a cogent overview of the processes and outcomes of interdisciplinary learning, focusing on the positive outcomes of developing more advanced epistemological beliefs, metacognition skills, and enhanced critical-thinking ability. They apply Biggs and Collis's (1982) SOLO (structure of observed learning outcomes) Model to interdisciplinary learning outcomes to reveal a progression from declarative and procedural knowledge, to interdisciplinary content thinking, and ultimately to a well-developed interdisciplinary knowledge structure with the ability to transfer to new interdisciplinary problems and themes.

Despite the rewards of interdisciplinary work, there are several challenges. For example, it is often difficult to integrate different disciplinary objects and methods of study, languages, and cultures (Schoenberger 2001). Disciplinary cultures are translated into epistemological commitments and self-reproducing social orders that can pose substantial barriers in interdisciplinary research and teaching. A genuine and productive interdisciplinary engagement is only possible when different disciplinary cultures are understood and worked through. Schoenberger (2001) warns against several other pitfalls common in interdisciplinary work. One is disciplinary reductionism where disciplinary knowledge is condensed into a set of facts and data that may oversimplify and stereotype a discipline. Another is disciplinary imperialism where one discipline extracts information from another discipline but does not value the discipline as a whole. Such practices prevent shared understanding and may impede collaborative efforts (Graybill et al. 2006).

Clear communication is essential in interdisciplinary contexts, although it can be difficult because of the different nuances, connotations, and denotations of terms across disciplines (Brown 1954). To avoid this problem and create shared intellectual understanding, interdisciplinary projects must allocate sufficient time at the beginning of a collaboration to develop shared vocabularies (Bracken and Oughton 2006). Not only do core concepts and terminology differ across disciplines, but also in many cases so do the concepts of what constitutes valid evidence and methods. Without adequate time to communicate and share expectations, breakdown of collaboration may occur (Rhoten and Parker 2004; Graybill et al. 2006). Language may also pose a problem externally through the process of evaluation and peer review, making acceptance of interdisciplinary papers by highly rated journals difficult and affecting the career advancement of junior faculty (Marts 2002).

A FEW FINAL WORDS AS YOU PONDER YOUR EXCITING FUTURE CAREER CHOICES

For graduate students especially, we urge you most strongly to read the article "A Rough Guide to Interdisciplinarity: Graduate Student Perspective" by a cohort of urban ecology graduate students about their experience in an IGERT program (Graybill et al. 2006). The article offers a view from the graduate student perspective on the stressors and rewards of interdisciplinary training and research. We believe their report will be of value to you as you consider the numerous emerging possibilities of pursuing interdisciplinary work in your doctoral program and future academic career.

References

Alberti, M., J. Marzluff, G. Bradley, C. Ryan, E. Shulenberger, and C. ZumBrunnen. 2003. Integrating humans into ecology: Opportunities and challenges for studying urban ecosystems. *Bioscience* 53: 1169–79.

Bellinger, G. 1994. Mental model musings. http://www.systems-thinking.org/index.htm.

Biggs, J. B., and K. F Collis, eds. 1982. *Evaluating the quality of learning: The SOLO taxonomy (Structure of the Observed Learning Outcome).* New York: Academic Press.

Bracken, L. J., and E. A. Oughton. 2006. "What do you mean?" The importance of language in developing interdisciplinary research. *Transactions of the Institute of British Geographers* 31:371–82.

Campbell, L. A. 2005. Overcoming obstacles to interdisciplinary research. *Conservation Biology* 19:574–77.

Caviglia-Harris, J. L., and J. Hatley. 2004. Interdisciplinary teaching: Analyzing consensus and conflict in environmental studies. *International Journal of Sustainability in Higher Education* 5:395–403.

Chappell, A. 2006. Using the "grieving" process and learning journals to evaluate students' responses to problem-based learning in an Undergraduate Geography Curriculum. *Journal of Geography in Higher Education* 30:15–31.

Committee on Science, Engineering, and Public Policy (COSEPUP). 1995. *Reshaping the graduate education of scientists and engineers.* Washington, DC: National Academy Press.

Dubrow, G., and J. Harris. 2006. Seeding, supporting, and sustaining interdisciplinary initiatives at the University of Washington: Findings, recommendations, and strategies, Draft Report for The Graduate School, Jan 17, 2006, University of Washington, Seattle, WA. http://www.grad.washington.edu/Acad/interdisc_network/ID_Docs/Dubrow_Harris_Report.pdf (last accessed 21 September 2007).

Duch, B., S. Groh, and D. Allen, eds. 2001. *The power of problem-based learning: A practical "how to" for teaching undergraduate courses in any discipline.* Sterling, VA: Stylus Publishing.

Farley, J., J. Erickson, and H. Daly. 2005. *Ecological economics: A workbook for problem-based learning.* Washington, DC: Island Press.

Golde, C. M., and T. M. Dore. 2001. *At cross purposes: What the experiences of doctoral students reveal about doctoral education.* http://www.phd-survey.org (last accessed 21 September 2007).

Goldstein, R. E., K. Visscher, R. Reinking, L. A. Oland, and M. Tabor. 2004. An interdisciplinary graduate laboratory for biological physics. PACS numbers: 87.16.Nn, 87.17.Nn, 87.17.Jj, 87.18.Pj (25 October 2004).

Graybill, J., S. Dooling, V. Shandas, J. Withey, A. Greve, and G. L. Simon. 2006. A rough guide to interdisciplinarity: Graduate student perspective. *BioScience* 56 (9): 757–63. http://www.biosciencemag.org (last accessed 21 September 2007).

Harris, D. L., R. C. Henry, C. J. Bland, S. M. Starnaman, and K. L. Voytek. 2003. Lessons learned from implementing multidisciplinary health professions educational models in community settings. *Journal of Interprofessional Care* 17:7–20.

Holmes, J. H. 2002. Geography's emerging cross-disciplinary links: Process, causes, outcomes and challenges. *Australian Geographical Studies* 40:2–20.

Ivanitskaya, L., D. Clark, G. Montgomery, and R. Primeau. 2002. Interdisciplinary learning: Process and outcomes. *Innovative Higher Education* 27:95–111.

Klein, J. T. 1996. *Crossing boundaries: Knowledge, disciplinarities, and interdisciplinarities.* Charlottesville, VA: University Press of Virginia.

———. 1999. *Mapping interdisciplinary studies.* Washington, DC: Association of American Colleges and Universities.

———. 2004. Prospects for transdisciplinarity. *Futures* 36:515–26.

———. 2005. *Humanities, culture, and interdisciplinarity: The changing American academy.* Albany: State University of New York Press.

Lattuca, L. 2001. *Creating interdisciplinarity: Interdisciplinary research and teaching among college and university faculty.* Nashville: Vanderbilt University Press.

Lélé, S., and N. B. Norgaard. 2004. Practicing interdisciplinarity. *BioScience* 55:967–5.

Manathunga, C., P. Lant, and G. Mellick. 2006. Imagining an interdisciplinary doctoral pedagogy. *Teaching in Higher Education* 11:365–79.

Marts, S. A. 2002. Interdisciplinary research is key to understanding sex differences: Report from the Society for Women's Health Research meeting on understanding the biology of sex differences. *Journal of Women's Health and Gender-Based Medicine* 11:501–9.

Monk, J., P. Manning, and C. Denman. 2003. Working together: Feminist perspectives on collaborative research and action. *ACME: An International E-Journal for Critical Geographies* 2:91–106.

National Research Council. 1994. *Federal support of basic research in institutions of higher learning.* Washington, DC: National Academy of Sciences.

Nerad, M., and J. Cerny. 1999. *PhDs: 10 years later study.* www.educ.washington.edu/COEWebsite/Cirge/HTML/research_projects.html.

Nyquist, J. 2000. *Re-envisioning the Ph.D. to meet the needs of the 21st century.* http://www.grad.washington.edu/envision/practices/index.html (last accessed 21 September 2007).

Nyquist, J., L. Manning, and D. H.Wulffe. 1999. On the road to becoming a professor: The graduate student experience. *Change: The Magazine of Higher Learning* 31:18–27.

Pallas, A. M. 2001. Preparing education doctoral students for epistemological diversity. *Educational Researcher* 30:6–11.

Pawson, E. 2006. Problem-based learning in geography: Towards a critical assessment of its purposes, benefits, and risks. *Journal of Geography in Higher Education* 30:103–16.

Reich, S. M., and J. A. Reich. 2006. Cultural competence in interdisciplinary collaborations: A method for respecting diversity in research partnerships. *American Journal of Community Psychology* 38:51–62.

Rhoten, D., and A. Parker. 2004. Risks and rewards of an interdisciplinary research path. *Science* 306:20–46.

Rossini, F. A., and A. L. Porter. 1984. Interdisciplinary research: Performance and policy issues. In *Problems in interdisciplinary studies*, eds. R. Jurkovich and J. H. P. Paelinck, 26–45. Brookfield, VT: Gower Publishing.

Schoenberger, E. 2001. Interdisciplinarity and social power. *Progress in Human Geography* 25:365–82.

Scholtes, P. R., B. L. Joiner, and B. J. Streibel. 2003. *The team handbook,* 3rd ed. Madison, WI: Oriel Incorporated.

Semerjian, L., M. El-Fadel, R. Zurayk, and I. Nuwayhid. 2004. Interdisciplinary approach to environmental education. *Journal of Professional Issues in Engineering Education and Practice* 130:173–81.

Senge, P. M. 1990. *The fifth discipline: The art & practice of the learning organization.* New York: Doubleday Currency.

Shearer, M. C. 2007. Implementing a new interdisciplinary module: The challenges and the benefits of working across disciplines. *Practice and Evidence of the Scholarship of Teaching and Learning in Higher Education* 2:2–22.

Sonnert, G. 2002. *Ivory bridges: Connecting science and society.* Cambridge, MA: MIT Press.

Tress, B., G. Tress, and G. Fry. 2003. Potential and limitations of interdisciplinary and transdisciplinary landscape studies. In *Interdisciplinarity and transdisciplinarity landscape studies: Potential and limitations,* eds. B. Tress, G. Tress, A. van der Walk, and G. Fry, 182–91. Wageningen, The Netherlands: Delta Program.

White III, H. B. 2001. A PBL course that uses articles as problems, Chapter 12. In *The power of problem-based learning: A practical "how to" for teaching undergraduate courses in any discipline,* eds. B. Duch, S. Groh, and D. E. Allen, 131–140. Sterling, VA: Stylus Publishing.

INDEX